Hodder Education
338 Euston Road, London NW1 3BH.

Hodder Education is an Hachette UK company

First published in UK 2011 by Hodder Education.

This edition published 2011.

British Library Cataloguing in Publication Data: a catalogue record for this title is available from the British Library.

10 9 8 7 6 5 4 3 2 1

www.hoddereducation.co.uk

Typeset by MPS Limited, a Macmillan Company.
Printed in Great Britain by CPI Cox & Wyman, Reading.

Contents

1

starting physics

Physics has always been at the leading edge of human thought, not just in recent times but also before the Scientific Age, which began about four centuries ago. The theories of science that were held to be true long ago might seem very odd when we compare them with our present knowledge. We must not forget though that the scientists of ancient times, or natural philosophers as they were then called, came up with ideas and theories that were astonishingly sophisticated in comparison with other aspects of life in those times. In this chapter we will look at some of the physics ideas from long ago to see just how these ideas were developed and used. Then we will look at some of the important skills needed in physics, before moving on to use these skills to carry out density tests.

Before the Scientific Age

Many theories of science before the Scientific Age were based on the assumptions that the Earth was at the centre of the Universe and that living beings were created by one or more superior beings who designated a special role for humans.

Science as a recorded activity flourished in the culture of Ancient Greece and the Mediterranean civilizations that developed from Greece. The idea that matter is composed of '**atoms**' too small to be seen individually was a theory put forward by Democritus (470–400 BCE). Two centuries later, Aristarchus put forward the theory that the Sun was at the centre of the Universe. In the next century, Archimedes (287–212 BCE) made important discoveries in mechanics and mathematics. The importance of the scientific heritage bequeathed by these and other natural philosophers of the ancient Mediterranean civilizations is undisputed. Indeed the impact of one particular natural philosopher, Aristotle (382–322 BCE), was to influence the conduct of science for many centuries, right through to the beginning of the Scientific Age.

Aristotle shaped science into a coherent set of ideas that were consistent with the prevailing world-view, namely that the Earth is the centre of the Universe and therefore a special place has been accorded to the human race by the creator of the Universe. However, Aristotle's rejection of theories that did not fit in meant that interesting ideas were not followed up and investigated further. Aristotle stamped his method on science so firmly that it lasted for over 15 centuries. His approach of picking facts to support accepted theories dominated the way science was conducted long after his death, from Ancient Greece, through the Roman Empire, the Dark Ages and into the Middle Ages.

Archimedes, the first great scientist

Archimedes was born in Sicily, where he worked under the patronage of King Hiero, the island's ruler. Archimedes kept in touch with the scholars of Alexandria in Egypt, where one of his inventions, the water screw, was put to practical use to raise water from the river Nile. This device consists of a tight-fitting screw in a cylinder. With the cylinder in water at its lower end, turning the screw raises water up the cylinder from the lower end. Archimedes' screw allowed farmers to irrigate the land bordering the river Nile.

Archimedes also discovered the principle of the lever and worked out how a **force** could be increased using a lever. No doubt you have used the lever principle to move an object that is difficult to shift. For example, when you use a bottle opener as a lever to remove the cap from a bottle, the force you apply to the bottle opener causes a much larger force to act on the cap. The applied force on the lever is referred to as the effort. The lever pivots about a fixed point or fulcrum as shown in Figure 1.1. The force of the lever on the cap is much greater than the applied force because it acts much nearer to the fulcrum than the applied force does.

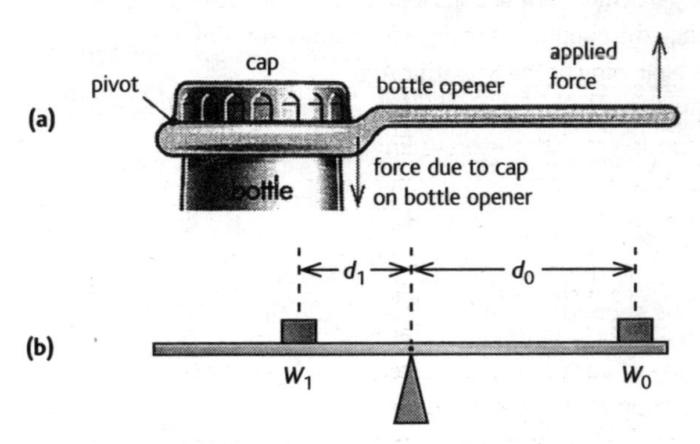

Figure 1.1 *(a) Using a lever. (b) Balancing a beam.*

Archimedes investigated the equilibrium conditions for a beam balanced at its centre of gravity with a **weight** supported on each side of the beam. He found that the distance of each weight to the fulcrum was in inverse proportion to the amount of weight.

* For two objects of equal weight, the distances from the fulcrum must be equal for the beam to be in equilibrium.

* For two objects of weights W_1 and W_0, the distances d_1 and d_0 of the weights from the fulcrum are in inverse ratio to the weights. In other words:

$$\frac{d_0}{d_1} = \frac{W_1}{W_0}$$

If one of the objects is of known weight, by measuring the distances d_1 and d_0 the other weight can be calculated using the equation above. For example, suppose one of the objects is a standard weight at a distance of 0.50 m from the fulcrum; if the beam is balanced by an object of unknown weight at a distance of 0.25 m from the fulcrum, the unknown weight must be exactly twice as heavy as the standard weight.

Weights and measures

The scientific system of units is known as the **SI system**. This was agreed internationally several decades ago. Before this agreement was reached, different systems of units were used in different countries. In earlier times, before national systems were established, local systems developed by traders varied according to custom and practice.

The five base units of the SI system are:

1. the **second**, which is defined as the time taken for a specified number of vibrations of a certain type of atom in an atomic clock.
2. the **metre**, which is defined as the distance travelled by light in a vacuum in a specified time.
3. the **kilogram**, which is the unit of mass. This is defined as the quantity of matter in a block of platinum kept in the Bureau of International Weights and Measures (BIPM) in Paris. Standard masses kept in other scientific laboratories are measured by comparison with the standard one kilogram mass in Paris. The weight of an object is the force of gravity on the object. Because the weight of an object is in

proportion to its mass, a beam balance as described earlier may be used to compare the masses of any two objects.

4 the **kelvin**, which is the scientific unit of temperature.

5 the **ampere**, which is the scientific unit of electric current.

All other scientific units are derived from these five base units. Some examples of derived units are given in the following table. Note that in writing the symbol for a unit, the historical use of the oblique symbol '/' for 'per' (as in metre/second) is usually replaced in more advanced work by a numerical superscript after the symbol (as in ms^{-1}).

Quantity	*Definition*	*Unit*
Area of a rectangle	length × breadth	square metres, m^2
Volume of a box	length × breadth × height	cubic metres, m^3
Density	mass per unit volume	kilogram per cubic metre, kg/m^3 or $kg\ m^{-3}$
Speed	distance moved per unit of time	metre per second, m/s or ms^{-1}

Powers of ten

Scientific measurements and calculations involve values that can range from extremely small to enormously large. For example, the diameter of an atom is about 0.0000000003 metres and the distance from the Earth to the Sun is about 150 000 000 000 metres. Such values are usually written in **standard form** as a number between 1 and 10 multiplied by an appropriate power of ten. For example, the distance from the Earth to the Sun written in this way is 1.5×10^{11} m, where m is the abbreviation for metres and $10^{11} = 100\ 000\ 000\ 000$. Note that 10^{11} is quoted as 'ten to the power eleven'. The above value for the diameter of an atom would be written as 3.0×10^{-10} m, where $10^{-10} = 1 \div 10^{10} = 0.0000000001$. Note that a negative power of ten is a code for expressing powers of ten less than 1. Thus 10^{-10} is 1 divided by 'ten to the power ten', usually quoted as 'ten to the power minus ten'.

Scientific prefixes are used to represent certain powers of ten. For example, the prefix kilo- represents a thousand (= 10^3). Other standard scientific prefixes are listed below.

Prefix	Tera	Giga	Mega	kilo	milli	micro	nano	pico
Prefix symbol	T	G	M	k	m	μ	n	p
Power of ten	10^{12}	10^9	10^6	10^3	10^{-3}	10^{-6}	10^{-9}	10^{-12}

Notes

1 The prefix symbol for micro, μ, is pronounced 'mu'.
2 1 centimetre (cm) = 10^{-2} m and 1000 grams = 1 kilogram.

Density tests

The **density** of a substance is its **mass** per unit volume. The scientific unit of density is the kilogram per cubic metre (kg/m^3).

For example, the density of lead is about 11 000 kg/m^3. In comparison, the density of water is 1000 kg/m^3. Thus lead is 11 times more dense than water. The mass of a certain volume of lead is 11 times greater than the mass of the same volume of water.

Density calculations

$$\text{Density} = \frac{\text{mass}}{\text{volume}}$$

The unit of density is the kilogram per cubic metre (kg/m^3).

The density formula can be rearranged as:

$$\text{Mass} = \text{volume} \times \text{density, or volume} = \frac{\text{mass}}{\text{density}}$$

Density measurements – liquids

* The volume of a liquid is measured by pouring the liquid into an empty measuring cylinder and measuring the level of the liquid in the cylinder against its graduated scale. The scale is usually marked in cubic centimetres (cm^3). To convert the

reading into cubic metres (m^3), divide the measurement in cm^3 by 1 million (= 10^6) because $1\ m^3 = 10^6\ cm^3$.

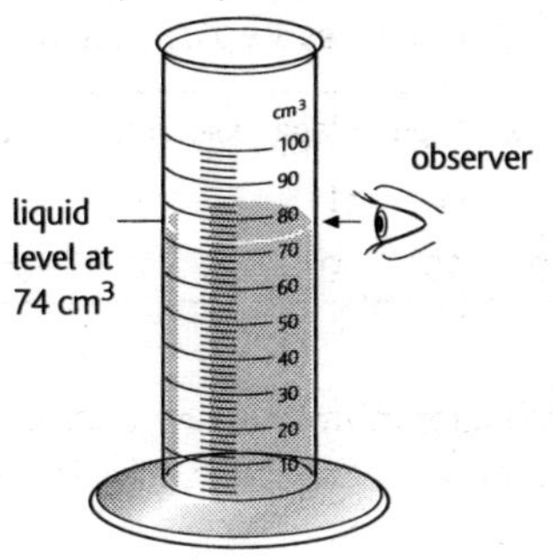

Figure 1.2 *Using a measuring cylinder.*

* The mass of a measured volume of liquid can be measured by using a top pan balance. This is used to measure the mass of an empty beaker. Then the liquid is poured into the beaker and the total mass of the beaker and the liquid is remeasured. The mass of the liquid is the difference between the total mass of the beaker and liquid and the mass of the empty beaker. A top pan balance usually gives a reading in grams (g). Note that 1000 g = 1 kg.

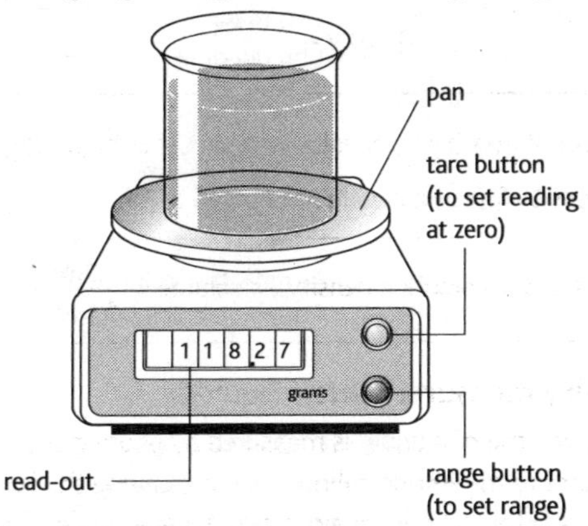

Figure 1.3 *Using a top pan balance.*

* The density is calculated by dividing the mass in kilograms by the volume in m^3.

Sample measurements

Volume of liquid in measuring cylinder = 90 cm^3 = 90×10^{-6} m^3
Mass of empty beaker = 115 g
Mass of beaker and liquid from measuring cylinder = 220 g
Therefore, the mass of liquid = 220 – 115 = 105 g = 0.105 kg

$$\text{Density of liquid} = \frac{0.105\ \text{kg}}{90 \times 10^{-6}\ \text{m}^3} = 1170\ \text{kg/m}^3$$

Density measurements – solids

* Use a top pan balance to measure the mass of a piece of a solid.
* Pour some water into a suitable measuring cylinder. Note the volume of the water from the reading of the water level. Tie the piece of solid to a thread and lower it into the water until it is completely submerged. The level of water in the cylinder will rise as a result. Note the new reading of the water level. The volume of the piece of solid is the difference between the two readings.
* The density is calculated by dividing the mass in kilograms by the volume in m^3.

Notes

1 The solid must be insoluble in water.
2 The volume of a regular solid such as a cube can be calculated from its dimensions. For example, the volume of a rectangular box is equal to its length × its height × its width.

Sample measurements

Mass of object = 35.4 g = 0.0354 kg
Volume of water only in measuring cylinder = 50 cm^3
Volume of water and object in measuring cylinder = 63 cm^3

Therefore, the volume of the object = $63 - 50 = 13\ \text{cm}^3 = 13 \times 10^{-6}\ \text{m}^3$

$$\text{Density of object} = \text{mass/volume} = \frac{0.0354\ \text{kg}}{13 \times 10^{-6}\ \text{m}^3} = 2720\ \text{kg/m}^3$$

Eureka!

Perhaps the most famous story in science is about Archimedes when he was asked by his king to find out if his new crown really was made of gold. The king thought the royal crown-maker might have cheated by making cavities in the gold, which could have been filled with lead. He did not want to cut the new crown up, so he asked Archimedes to solve the problem without doing so. The solution came to Archimedes when he got into a bath of water and observed that the level of the water rose as he lowered himself into it. He realized that when an object is lowered completely into water, the displacement of the water is a measure of the volume of the object. According to legend, he greeted this discovery by running naked through the streets shouting 'Eureka!', which means 'I have found it'.

So how did Archimedes use this discovery to find out whether the crown was a fake? Archimedes recognized that, if the crown was 100% gold, its weight in air divided by its volume should be the same as for any other gold object. So he weighed the crown and then measured its volume by measuring the volume of water displaced when it was immersed completely in water. He then calculated the weight of the crown in air divided by its volume. Then he repeated the test and calculation with an object known to be made of gold. Fortunately for the royal crown-maker, the result was the same. Archimedes' measurements showed that the crown's density was the same as the density of a solid gold object. Gold is 18 times more dense than water. In effect, Archimedes discovered that the crown and a known gold object were both 18 times as dense as water. He therefore concluded that the crown was made of gold. Archimedes' test is the first recorded example of non-destructive testing.

2

science in motion

The Scientific Age in which we live began uneasily several centuries ago. Perhaps the discovery of the New World made some scientists and philosophers think about nature differently from those who accepted Aristotle's teachings. Galileo is generally considered to be the founder of modern science, as he demonstrated that Aristotle's way of doing science was fundamentally flawed because it ignored the crucial role of observations. Over succeeding centuries, Galileo's approach to science was developed and the link between laws and observations became evident as new discoveries were made, analysed and tested. The modern philosophy of science is that laws and theories hold true on the basis of never having been disproved. This philosophy was formulated by Sir Karl Popper, who realized that observations and measurements can never prove a theory or a law but, at any time, a single experiment is sufficient to disprove a theory or law.

Galileo's experiments on motion

In a legendary demonstration at Pisa in Italy, Galileo was reported to have shown that two falling objects descend at the same rate, regardless of their weight. This could have been done by releasing two objects simultaneously from the top of the Leaning Tower and observing that they hit the ground at the same time, thus proving that they fall at the same rate.

Galileo wanted to find out how the time of descent would change if the distance fallen was changed. With falling objects, he was unable to measure the time of descent accurately. However, he realized that the motion of an object released at the top of an inclined plane was likely to be similar to that of a falling object but slow enough to time. He devised a water clock to measure the time taken by an object to reach certain distances down the plane. Pendulum clocks and other mechanical timing devices were inventions of the future. Galileo measured the amount of water running at a constant rate out of a tank from the moment the object was released at the top of the inclined plane to when it passed a certain marker. The amount of water collected thus served as a measure of the time of descent.

Motion in a straight line at constant speed

Speed is defined as distance travelled per unit of time. The scientific unit of speed is the metre per second (abbreviated as m/s or ms^{-1}).

For an object moving at constant speed, the distance travelled by the object in a certain time can be calculated by multiplying the speed by the time taken.

Distance = speed × time taken

For example, suppose a car is travelling at a constant speed of 12 m/s. The car therefore travels a distance of 12 metres each second and would travel a distance of 720 metres in 60 seconds

(= 12 m/s × 60 seconds). In 1 hour, which is 3600 seconds, the car would travel a distance of 43 200 metres (= 12 m/s × 3600 seconds), or 43.2 kilometres.

For an object moving at varying speed, its average speed over a certain distance or time is defined as the distance travelled/the time taken.

$$\text{Average speed} = \frac{\text{distance travelled}}{\text{time taken}}$$

Equations and symbols

Because word equations such as 'distance = speed × time' become tedious and time-consuming, letters are used as 'shorthand' symbols for physical quantities. So 'distance = speed × time' is usually written as $s = vt$, where s represents distance moved, v represents speed and t represents time taken. Each symbol stands for a certain physical quantity, in accordance with an agreed scientific convention.

An equation may need to be rearranged by moving the symbols from one side to the other. The basic rule to remember is to do the same to both sides of the equation.

For example, in the equation $s = vt$, the symbol s is said to be the subject of the equation because it appears first as you read the equation from left to right. To make v the subject of the equation,

* divide both sides of $s = vt$ by t, to give $\frac{s}{t} = \frac{vt}{t}$
* cancel t from the top and bottom of the right-hand side of the equation to give $\frac{s}{t} = v$
* swap the two sides of the equation over so v is read first; $v = \frac{s}{t}$

Velocity and speed

Velocity is defined as speed in a specified direction. The unit of velocity is the metre per second (m/s), the same as the unit of speed.

Two objects travelling at 60 mph in opposite directions along a motorway have the same speed but not the same velocity.

An object moving along a circular path at constant speed has a continually changing direction of motion, so its velocity continually changes even though its speed is constant.

Accelerated motion along a straight line

Acceleration is defined as change of velocity per unit of time. The unit of acceleration is the metre per second per second, usually written as m/s^2.

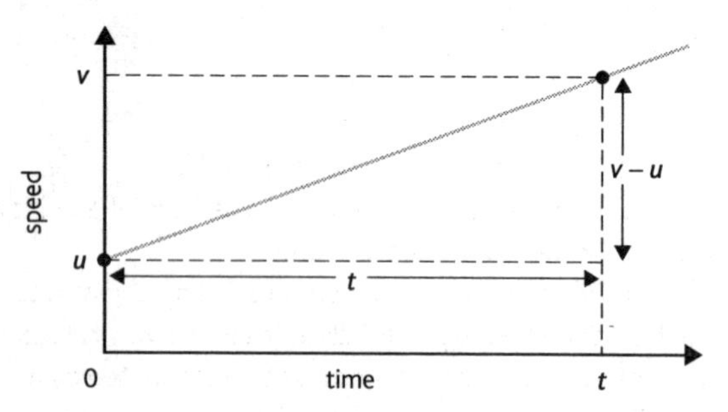

Figure 2.1 *A speed v. time graph for constant acceleration.*

For an object moving in one direction along a straight line, its acceleration is constant if its speed changes by equal amounts in equal times. Suppose the speed of such an object changes from u to v in time t. This change is shown by the straight line on the speed–time graph in Figure 2.1.

The change of speed in time $t = v - u$

Therefore, the acceleration, $a = \dfrac{\text{change of speed}}{\text{time taken}} = \dfrac{v-u}{t}$

Note that a negative value of acceleration is referred to as **deceleration**. This is the term used for slowing down.

Worked example

A car accelerates, without change of direction, from rest to a speed of 30 m/s in 60 s. Calculate its acceleration.

Solution

Initial speed $u = 0$, final speed $v = 30$ m/s, time taken $t = 60$ s

Therefore $a = \dfrac{v-u}{t} = \dfrac{30-0}{60} = 0.5 \text{ m/s}^2$

Galileo and gravity

Galileo showed that an object rolling down an inclined plane gathered speed at a constant rate as it descended. He showed that the acceleration was constant and could be made larger by increasing the steepness of the incline.

He thought that the acceleration of an object falling freely ought to be constant and ought to be the same for any object. However, he was unable to time the vertical descent of an object because his clock was not accurate enough.

One method of timing the descent of a falling object is to photograph a small steel ball as it falls in front of a vertical scale. If the apparatus is in a darkened room that is illuminated by a flashing stroboscope, an image of the ball is captured on the photograph each time the light flashes. Provided the light flashes at a constant rate, the photograph therefore shows the position of the ball at equal intervals. A video camera could be used instead of a film camera but the stroboscope would still be needed. If the flash rate of the stroboscope is known, the time interval between successive flashes can be worked out. Figure 2.2 shows the idea. Measurements from such an investigation are shown below.

Time from start (s)	0.0	0.1	0.2	0.3	0.4	0.5
Distance fallen (m)	0.0	0.05	0.20	0.45	0.80	1.25
Average speed (m/s)	0.0	0.50	1.00	1.50	2.00	2.50
Speed (m/s)	0.0	1.00	2.00	3.00	4.00	5.00

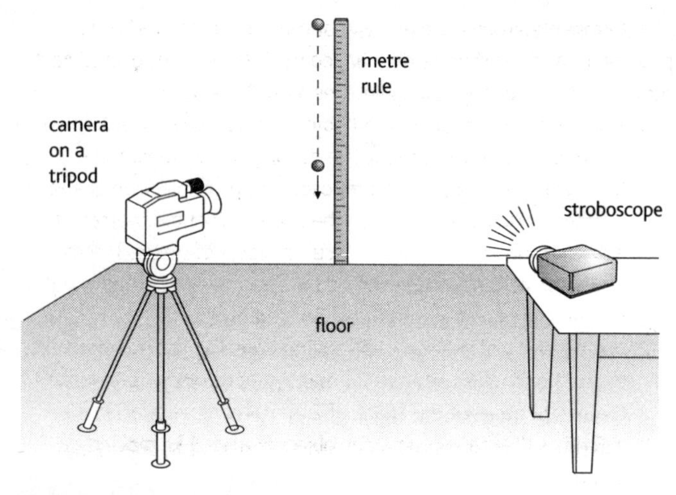

Figure 2.2 *Investigating free fall.*

The average speed between the start and each flash has been worked out by dividing the distance fallen by the time taken from the start. Because the initial speed is zero, the speed at each flash is twice the average speed. The results show that the speed increases by 1.0 m/s every 0.1 s, which scales up to an increase of 10 m/s every second. Thus the acceleration of free fall has a constant value of 10 m/s^2. This value is usually referred to as g, the acceleration due to gravity of a freely falling object. Accurate measurement of g shows that g varies slightly with latitude. Its value is 9.81 m/s^2 at the Earth's poles and 9.78 m/s^2 at the Equator.

Newton's laws of motion

Galileo's ideas on motion were developed by Sir Isaac Newton in his laws of motion and **gravitation**. For over two centuries, these laws were thought to be universal, applicable in any situation where material bodies interact. Newton's three laws of motion underpin much of modern technology and engineering. The motion of planets, comets and satellites can be explained almost completely using Newton's theory of gravitation and his laws of motion.

Before Newton wrote his laws of motion, he defined key physical quantities that he then linked in his laws of motion to each other and to other physical quantities already defined.

1 The quantity of matter in an object, or its **mass**, was defined by Newton as its volume × the density of the substance from which the object is composed. For example, iron is about 8 times as dense as water, so the mass of 2 cubic metres of iron is 16 times the mass of 1 cubic metre of water. The mass of an object is now determined by comparing its weight with the weight of a standard mass. See Chapter 1.
2 The quantity of motion of a moving object, or its **momentum**, was defined by Newton as its mass × its velocity. For example, the momentum of an object of 2 kg mass moving at 3 m/s is 6 times the momentum of an object of mass 1 kg moving at 1 m/s.
3 A force is an action on an object that changes its motion. Two or more forces acting on an object balance each other out if the object stays at rest or moves at constant speed along a straight line.

Newton's 1st law of motion

An object continues at rest or in uniform motion unless acted on by a force.

Here are some everyday situations that reveal the link between force and motion as established by Newton in his 1st Law:

1 An object sliding across ice moves at constant speed without change of direction because, due to the ice, there is no force acting on the object. If the object crosses a surface where there is no ice, it is brought to a standstill by the force of friction between the object and the surface.
2 A cyclist in motion on a level road gradually comes to a halt if he or she stops cycling and 'freewheels'. The reason is that friction in the wheel bearings and air resistance on the cyclist act against the motion and gradually stop the cycle moving.

In effect, Newton's 1st Law tells us what a force is – namely, anything that can change the motion of an object. Different types of force include:

* the force of gravity on an object, usually referred to as the weight of the object
* the force of friction acting on two surfaces when they slide over each other
* the tension in an object such as a string when it is taut and being pulled at each end
* the force on an object when another object pushes or pulls it or supports it
* the force between two electrically charged objects or between two magnetized objects.

Newton's 2nd law of motion

The force on an object is proportional to the rate of change of momentum of the object.

An object initially at rest and acted on by a single force gains speed and therefore gains momentum. For a given amount of force, the object gains a certain amount of momentum every second during the time the force acts on it. Consider the example of an object of mass m being accelerated along a straight line by a constant force F, acting in the same direction as the object is moving, for a time t. Suppose the object's speed increases from u to v in this time, without change of direction.

Force and motion

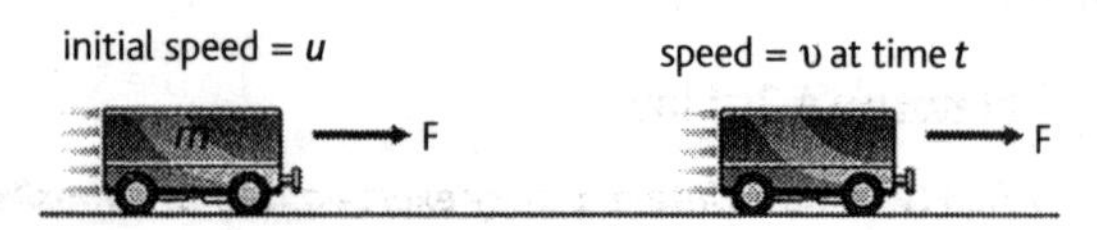

Figure 2.3 *Newton's 2nd Law of motion.*

* The initial momentum of the object = mu
* The final momentum of the object = mv

so the change of momentum = final momentum – initial momentum:

$$mv - mu$$

and the rate of change of momentum = $\frac{\text{Change of momentum}}{\text{time taken}}$:

$$\frac{mv - mu}{t} = \frac{m(v - u)}{t}$$

According to Newton's 2nd Law, the force F on the object is proportional to the rate of change of momentum.

Hence:

$F = km\frac{(v-u)}{t}$ where k is a constant of proportionality.

Because the acceleration of the object $a = \frac{(v-u)}{t}$, then $F = kma$.

By defining the unit of force, the **newton**, as the amount of force that would give an object of mass 1 kg an acceleration of 1 m/s^2, the value of k is therefore set at 1. Newton's 2nd Law can then be written as the equation

$$F = ma$$

where F = force in newtons, m = mass in kilograms, and a = acceleration in m/s^2.

Newton's 3rd law

When two bodies interact, they exert equal and opposite forces on each other.

For example, if you clap your hands together, the force of your left hand on your right hand is equal and opposite to the force of your right hand on your left hand.

If you lean on a wall, the wall exerts a force on you that is equal and opposite to the force you exert on the wall. When you stand still, the floor exerts an upward force on you that is equal and opposite to the force you exert on the floor.

Mass and weight

The acceleration of a freely falling object, g, is constant, provided the distance fallen is much smaller than the radius of the Earth. This acceleration is caused by the weight of the object, which is the force of gravity pulling it towards the Earth. Using Newton's 2nd Law, in the form $F = ma$, therefore gives mg as the force of gravity on an object of mass m. Hence the weight W of an object of mass $m = mg$. Weight W (in newtons) $= mg$, where m = mass in kilograms and g = the acceleration of a freely falling object ($= 9.8 \text{ m/s}^2$ near the Earth's surface). The weight of a 1 kg mass near the Earth's surface is therefore 9.8 N.

The weight of a person of mass 60 kg at the Earth's surface is 588 N. **The gravitational field strength** at any position is defined as the force of gravity per unit mass on an object at that position. Hence gravitational field strength $= mg/m = g$. Thus g can be referred to as either the acceleration of a freely falling object (in m/s^2) or gravitational field strength (in newtons per kilogram).

Stability

Force as a vector

A physical quantity that has a direction as well as a size is a **vector** quantity. Examples of vectors include force, velocity, acceleration and momentum. A physical quantity that is not directional is referred to as a **scalar** quantity. Examples of scalars include distance, speed, mass and energy.

A vector quantity may be represented by an arrow of length in proportion to the quantity and pointing in the appropriate direction. For example, a force of 10 N acting horizontally due North on an object may be represented on a diagram as an arrow of length 10 cm pointing in a direction defined on the diagram as due

North. Caution is needed to use Newton's 3rd Law correctly when force diagrams are drawn. For every force acting on a body, there is an equal and opposite force exerted by the body. To ensure that force diagrams do not become complicated, a force diagram for an object should show only the forces acting on the body, not the forces exerted by the body.

Centre of gravity

When an object is acted upon by two or more forces including the force of gravity, the effect on the object depends on the direction and size of the forces. Each force acting on an object may be represented as a vector on a force diagram. To represent the force of gravity on an object (i.e. its weight), which acts on all parts of the object, we define the centre of gravity of an object as the point where its entire weight may be considered to act. The weight of an object can then be shown on a force diagram as a vector arrow pointing downwards at the centre of gravity.

The idea of centre of gravity, sometimes referred to as centre of mass, enables us to understand stability and the conditions necessary to keep an object in equilibrium.

For example a tall free-standing bookcase is liable to topple over when it is moved, unless it is supported. If it tilts too much, its centre of gravity moves outside its base and it topples over, unless someone supports it to prevent this. Push a tall object such as an empty cereal packet sideways at its top and you will find that it tilts, then topples over if it is pushed too far. If you hold it at the point where it only just topples, you should find the centre of the box is directly above the edge in contact with the table.

3

energy

The industrial revolutions in the eighteenth and nineteenth centuries developed because engineers were able to make powerful machines and engines that could do far more work than the windmills and watermills of the pre-industrial era. The scientific principles established by Newton enabled engineers to design and build such machines and engines. With Newton's guiding principles, engineers continue to develop machines that use energy more efficiently. In this chapter we look at what we mean by energy, power and efficiency.

The invention of the steam engine, the first powerful heat engine, started the industrial revolution of the nineteenth century. In the twentieth century, the internal combustion engine and the jet engine took over from the steam engine. In this chapter, we will also look at temperature, heat and the laws of thermodynamics, which set out the principles behind energy transformations.

Work and energy

The **work** done by a force on an object is defined as the force × the distance moved by the object along the line of action of the force. The unit of work is the joule (J), which is equal to the work done when a force of one newton moves its point of application through a distance of one metre in the direction of the force.

Work done (in joules)	=	force (in newtons)	×	distance moved in the direction of the force (in metres)

For example, the work done on a brick of weight 2.0 N lifted through a height of 1 m is equal to 2 J. If the same brick had been raised through only 0.5 m, the work done on it would have been 1 J.

The work done on an object by raising it enables the object to do work on something else when it drops. A brick dropped onto someone's shoe would do work in denting the shoe. The work done by the brick on the shoe could not be recovered, however, unless the shoe cap was elastic. A raised object has the capacity to do work because of its position. Another example is a clockwork spring, which has work done on it when it is wound up. When the spring unwinds, it does work by making the wheels and gears inside the clock move. If the spring is prevented from unwinding, it keeps the capacity to do work until it is allowed to unwind. This capacity to do work is defined as the **energy** of an object. The term 'stored work' might seem more appropriate, but work is only done when a force moves, so the term 'energy' is used for the capacity of a body to do work.

The energy of an object is its capacity to do work.

Energy is measured in joules, the same unit as the unit of work. For example, if a brick of weight 5 N is raised by 2 m from the ground, the work done to raise it is equal to 10 J (= 5 N × 2 m). The raised brick therefore contains an extra 10 J of energy due to being lifted. If the brick is dropped onto someone's shoe, it causes a dent in the shoe and loses the extra 10 J of energy in the process. Thus 10 J of work must have been done to create the dent. You can work out the force of the impact from the depth of the dent by recalling that work done = force × distance. For example, a dent of 0.5 cm (= 0.005 m) would correspond to a force of 2000 N (= 10 J/0.005 m) and a really big 'ouch'!

Energy transformations

Objects can possess energy in various ways, referred to as 'forms' of energy. For example:

* **Kinetic energy** is the energy of a moving object due to its motion. The faster an object moves, the more kinetic energy it has (see 'More about kinetic energy' later in this chapter).
* **Potential energy** is the energy of an object due to its position. For example, an object raised from the ground gains potential energy because a lift force opposite to the weight of the object is needed to raise the object.
* **Elastic energy** is the energy stored in an object when it is stretched or squeezed. When the object returns to its undistorted shape, the elastic energy is released.
* **Electrical energy** is the energy of particles such as **electrons** that carry electric charge.
* **Thermal energy** is the energy of an object due to its temperature. For example, in a car engine, the petrol–air mixture in a cylinder is ignited by means of a spark, which causes the air to become very hot very quickly. The **pressure** of the hot air forces the piston out so it turns the drive shaft.
* **Nuclear energy** is energy that can be released from an atom that has an unstable nucleus. See Chapter 7.
* **Chemical energy** is the energy that can be released when chemicals react. For example, when a battery is used to light a torch bulb, chemical energy is released inside the battery to force electricity through the torch bulb.

Energy can change from one form into other forms. We say that energy is 'transformed' when it changes from one form into another. For example, an object released above the ground loses potential energy and gains kinetic energy as it falls. So its gravitational potential energy is transformed into kinetic energy. If it hits the ground without bouncing, its kinetic energy is transformed into thermal energy and sound energy on impact.

The Principle of Conservation of Energy

When energy is transformed or transferred between objects in an isolated system, the total energy of all the objects after the change is the same as the total energy of all the objects before the change. The total energy of all the objects is conserved (i.e. unchanged). This is known as the Principle of Conservation of Energy. Energy is the capacity to do work. The Principle of Conservation of Energy means that the total capacity (to do work) of an isolated system of objects is unchanged by interactions between the objects. The principle has been tested many times and the measurements always come up with the result that the total energy is unchanged. When energy transfer takes places within the system, individual objects do work on each other, so gaining or losing energy. However, the total capacity to do work is unchanged. In other words, the total energy is unchanged.

Energy and power

Power is rate of transfer of energy.

The unit of power is the watt (W), which is equal to a rate of transfer of energy of 1 joule per second. A 3000-watt electric heater emits heat at a rate of 3000 joules every second. A 100-watt light bulb uses electrical energy at a rate of 100 joules per second. Every minute, a 100-watt light bulb would use 6000 joules of electrical energy. A weightlifter who raises an 800 N weight through a height of 1 metre in 2 seconds has a power output of 400 watts, because the weight is supplied with 800 J of potential energy (= weight × height gain) in 2 seconds.

$$\text{Power (in watts)} = \frac{\text{energy transferred (in joules)}}{\text{time taken (in seconds)}}$$

Efficiency and power

* The **power output** of a machine is the energy it supplies each second to the object it drives.
* The **power input** to a machine is the energy supplied to the machine each second.

Each second, a certain amount of energy is supplied to the machine and the machine delivers a certain amount of energy to whatever it is driving. The energy per second delivered by the machine cannot exceed the energy per second supplied to the machine, because energy cannot be created in the machine. If all the energy delivered by the machine each second is equal to the energy supplied to it each second, the machine is perfectly efficient. In practice, friction is usually present between the moving parts of a machine and so most machines are not perfectly efficient.

The **efficiency** of a machine is defined as

$$\frac{\text{the energy per second delivered by the machine}}{\text{the energy supplied to the machine per second}}$$

Because the energy per second delivered by the machine is its power output, and the energy per second supplied to it is its power input,

$$\text{Efficiency} = \frac{\text{power output}}{\text{power input}}$$

The maximum value of efficiency of a machine is 1. This is when the power output is equal to the power input and no power is wasted in the machine.

Note that the percentage efficiency of a machine is its efficiency × 100%.

More about kinetic energy

Kinetic energy is the energy possessed by a moving object due to its motion. Consider an object of mass m that accelerates from rest at constant acceleration a for a certain time t.

* The speed of the object after time t, v = acceleration × time = at.
* Hence the distance moved s = average speed × time = $\frac{1}{2} vt$ since the average speed = $\frac{1}{2} v$.
* Therefore the work done on the object = force × distance moved = (mass × acceleration) × distance moved $= m\,(v/t) \times \frac{1}{2} vt = \frac{1}{2} mv^2$
* Because all the work done goes to kinetic energy, it follows that the kinetic energy of the object at speed v is equal to $\frac{1}{2} mv^2$

Kinetic energy = $\frac{1}{2} mv^2$

Worked example

Calculate the kinetic energy of a vehicle of mass 800 kg moving at a speed of 30 m/s.

Solution

Kinetic energy $\frac{1}{2} mv^2 = \frac{1}{2} \times 800 \times 30^2 = 36\,000$ J.

Heat and temperature

In winter in Siberia, the temperature outdoors can fall below −40°C. In summer in the Sahara desert, the temperature can rise above 40°C. These temperatures are expressed on the Celsius scale, denoted by the symbol °C (degree C). This scale is defined in terms of two 'fixed points', which are:

* ice point, the melting point of pure ice at atmospheric pressure, 0°C
* steam point, the boiling point of pure water at atmospheric pressure, 100°C.

A fixed point is a standard 'degree of hotness' that can be reproduced as required. A thermometer calibrated in °C would read 0 when in pure melting ice at atmospheric pressure, and 100 when in steam at atmospheric pressure. Figure 3.1 shows a mercury-in-glass thermometer. The temperature of the mercury in the bulb is measured by reading where the end of the mercury column is against the scale. The thermometer is used by placing the bulb at the location where the temperature is to be measured, then reading the scale after allowing sufficient time for the reading to stop changing.

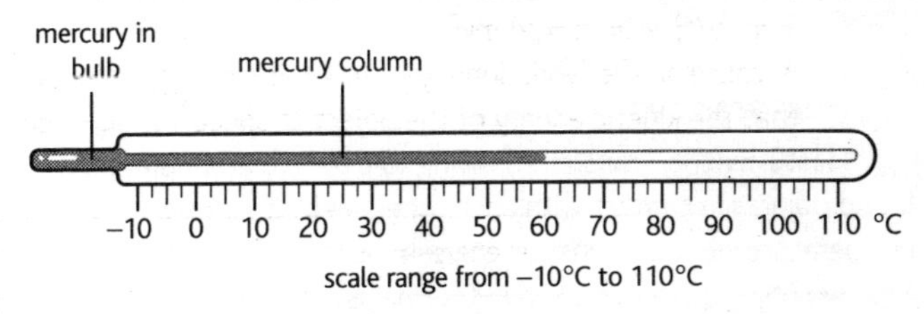

Figure 3.1 *A mercury-in-glass thermometer.*

The mercury thermometer works because the mercury in the bulb expands when it becomes warmer and moves along the air column in the glass tube. Mercury is a very suitable liquid for use in a thermometer because it expands evenly with increasing temperature. Coloured alcohol is used as the liquid in thermometers for low temperature measurements, because mercury freezes below −39°C.

Absolute zero

An object placed in a refrigerator is made colder by the refrigerator because the refrigerator removes energy from the object. How cold can an object be? The lowest temperature possible is called **absolute zero**. No energy can be obtained from an object at this temperature. The pressure of a gas would be zero at absolute zero, provided the gas did not liquefy as it was cooled. By measuring the pressure of a fixed amount of gas at steam point, then at ice point, the temperature for zero pressure (i.e. absolute zero) can be estimated. This temperature is −273°C. No matter which gas or how much gas is used, the same value is always obtained for absolute zero.

The ideas about absolute zero were developed by Lord Kelvin in the nineteenth century. The absolute scale of temperature in kelvins (K) is defined from two fixed points, which are:

* absolute zero at zero kelvins (= −273°C)
* the temperature at which ice, water and water vapour coexist (= 0°C).

Hence absolute temperature in kelvins = temperature in °C + 273.

Heat transfer

Heat is energy transferred owing to a temperature difference. If two objects in contact with each other are at the same temperature, no heat transfer between the two objects takes place. The two objects are said to be in thermal equilibrium.

Heat transfer can occur in three different ways:

1. **Thermal conduction** takes place in solids, liquids and gases when there is a temperature difference between different parts of the substance. Metals are the best conductors of heat for the same reason that they are good conductors of electricity; they contain tiny particles called electrons, which can move about carrying energy from one part of the metal to another part when there is a voltage or a temperature difference across the metal.
2. **Thermal convection** is the process of circulation that takes place in liquids and gases when there is a temperature difference between different parts of the liquid or gas. The density of a liquid or gas varies with temperature,

so temperature differences in a liquid or gas cause density variations. Gravity makes denser regions sink and less dense regions rise, thus causing circulation currents. A hot air balloon rises because the air in the balloon is heated by a burner and so becomes less dense and is therefore forced upwards.

3 **Thermal radiation** is emitted from every surface. The hotter the surface is, the greater the thermal radiation from it. No substance is needed to carry the radiation as it can travel through a vacuum.
   * A black surface is a much more effective absorber of thermal radiation than a shiny silvered surface. You can notice this if you hold a piece of black paper and a sheet of tinfoil at the same distance from a hot object. The black paper warms up much faster than the tinfoil.
   * A black surface is a much more effective radiator of thermal radiation than a shiny silvered surface. If you want to keep an object warm, wrap it in tinfoil so that it does not lose energy due to thermal radiation.

Thermal properties of materials

All substances are composed of molecules, which are the smallest particles of a substance. In solids, the molecules are held together rigidly by strong **bonds**. In liquids, the bonds are not strong enough to keep the molecules in a rigid structure. In a gas or a vapour, the molecules move about freely at large distances from each other. When energy is supplied to a substance to increase its thermal energy, the particles of the substance

* gain kinetic energy, if the temperature of the substance increases
* use the energy supplied to break the bonds between the molecules, if the substance changes state from a solid to a liquid or to a gas, or from a liquid to a gas.

Melting points and boiling points

When a pure solid is heated and heated, its temperature increases until it reaches the **melting point** of the solid. The solid changes to a liquid at this temperature, provided it continues to be heated. Continued heating raises the temperature of the liquid to its **boiling point**, the temperature at which the liquid boils. Freezing is the reverse process to melting and it occurs when a liquid has been cooled to its freezing point, which is the same temperature as the melting point. For example, the freezing point of water is 0°C, the same as the melting point of pure ice. The reverse process to boiling is condensation. If a gas in a container is cooled sufficiently, liquid forms on the surface of the container due to condensation.

Specific heat capacity

The **specific heat capacity** of a material is the energy needed to raise the temperature of unit mass of the material by one degree. For example, the specific heat capacity of water is 4200 joules per kilogram per °C. This means that:

* 4200 J of energy must be supplied to raise the temperature of 1 kilogram of water by 1°C
* 42 000 J of energy must be supplied to raise the temperature of 1 kilogram of water by 10°C
* 84 000 J of energy must be supplied to raise the temperature of 2 kilograms of water by 10°C.

More generally, to raise the temperature of mass m of a substance from T_1 to T_2,

$$\textit{the energy needed } \Delta E = mc\,(T_2 - T_1),$$

where c is the specific heat capacity of the material. The unit of c is J/kg/°C.

The laws of thermodynamics

The laws of thermodynamics are the principles behind energy transformations involving heat.

1 The **First Law of Thermodynamics** states that when an object or a system of objects does work or gains heat, the change

of **internal energy** of the object(s) is equal to the difference between the heat gained and the work done by the object(s).

* **Work** is energy transferred by means of a force when the force moves its point of application in the direction of the force. See 'Work, energy and power' earlier in this chapter.
* **Heat** is energy transferred due to a difference of temperature.
* The **internal energy** of an object is the energy it possesses, not counting its kinetic or potential or elastic energy. For example, if the temperature of a solid object is increased by heating the object, the internal energy of the object increases as its atoms vibrate more.

The First Law of Thermodynamics follows from the Principle of Conservation of Energy, as heat and work are the means by which an object gains or loses energy.

2 The **Second Law of Thermodynamics** states that energy tends to spread out and become less useful whenever it is transferred between objects in an isolated system.

For example, when an electric winch is used to raise a weight, energy is transferred from the electrical supply to the weight. However, some of the energy supplied is wasted due to friction between the moving parts and due to the heating effect of the electric current in the circuit. This wasted energy cannot be recovered and used to do work as it spreads out by means of heat transfer to the surroundings. The energy from the electrical supply that is gained by the weight is an example of useful energy. The energy from the electrical supply that is not gained by the weight is wasted as heat transfer to the surroundings.

4

electricity

Electricity provides most of the power we use at home and in the workplace. The principles that underpin electricity generation and distribution were discovered in the nineteenth century, and electricity distribution systems were set up in the late nineteenth century and the early decades of the twentieth century. In the future, renewable energy sources such as solar panels and wind turbines will probably provide electricity in urban and remote regions without causing pollution. In this chapter, we will look in depth at the principles of electricity and how electricity is generated.

The nature of electricity

Any material that allows electricity to pass through it is known as an **electrical conductor**. All metals and certain non-metals, such as graphite, conduct electricity. Materials that do not allow electricity to pass through them are known as **electrical insulators**. Examples of electrical insulators include polythene, nylon, Bakelite, air and oil. An electric torch consists of a battery connected to a torch bulb, usually with one terminal of the battery connected directly to the torch bulb and the other terminal connected to the torch bulb via a switch. When the switch is closed, the torch bulb lights because the two metal parts of the switch make contact. When the switch is opened, the two metal parts move apart and the bulb goes off. A complete circuit of electrical conductors is needed for electricity to pass through the torch bulb. If there is a gap in the circuit, no electricity can pass round the circuit.

Static electricity

Static electricity is produced in a thunderstorm when clouds become charged with electricity. When charged clouds can hold no more electricity, they discharge to Earth in massive lightning strokes. A less dramatic demonstration of static electricity can be achieved by rubbing an inflated balloon with a dry cloth. You may be able to feel tiny sparks of static electricity discharging to your hand if you then touch the charged balloon. The charged balloon will attract bits of paper and will stick to a ceiling. Certain insulating materials, such as glass, polythene, perspex, nylon and rubber, also become charged when rubbed with a dry cloth.

A charged object exerts a force on any other charged object. For example, two charged polythene rods repel each other when held close together. This may be demonstrated by charging the end of one of the rods and suspending it horizontally on a thread. The other rod is then charged at one end and repels the end of the suspended rod when placed close to it. The same effect is observed if two charged Perspex rulers are held close together. However, if a charged Perspex ruler is held near a charged polythene rod, the two objects attract each other. These tests show that:

1 there are two types of electric charge
2 two objects that carry like charge (i.e. the same type of charge) repel each other
3 two objects that carry unlike charge (i.e. different types of charge) attract each other.

The two types of charge are referred to as **positive** and **negative charge** because they cancel each other out if one type is brought into contact with the other type. A charged polythene rod carries a negative charge. Therefore, any charged object that is repelled by a charged polythene rod carries a negative charge. A charged Perspex ruler carries a positive charge and therefore it attracts a charged polythene rod.

The action of rubbing such a material charges it with electricity because tiny negatively charged particles transfer between the material and the cloth used to rub it. These charged particles are called **electrons** and they are in every atom of every substance. Every atom contains a positively charged nucleus surrounded by electrons. An uncharged atom has the same amount of negative charge as positive charge.

* If electrons are added to an uncharged atom, the atom becomes negatively charged because it now contains more negative charge than positive charge.
* If electrons are removed from an uncharged atom, the atom becomes positively charged because it now contains more positive charge than negative charge.

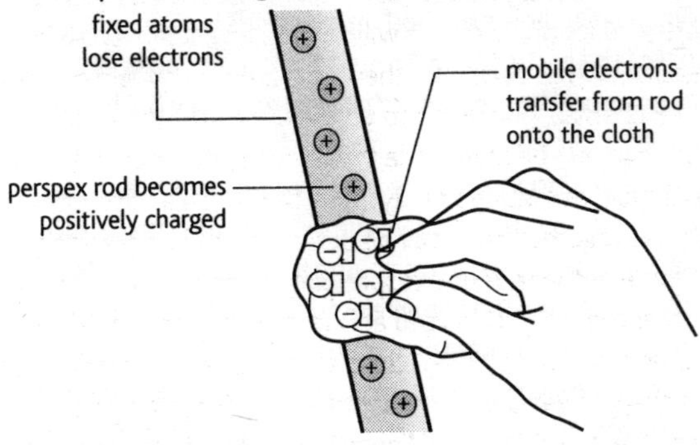

Figure 4.1 *Charging by friction.*

Current and charge

An electric current is a flow of charge, usually carried by electrons. The electrons in an insulator are trapped inside the atoms, but in a conductor some of the electrons are not trapped and can move about freely inside the conductor. A torch lamp circuit consists of a cell or a battery (two or more cells) connected to the torch bulb via two wires and a switch. When the switch is closed, electrons pass round the circuit because all the parts of the circuit can conduct electricity. The battery forces the free electrons in the conductors one way round the circuit through the wires, the torch bulb, the switch and the battery.

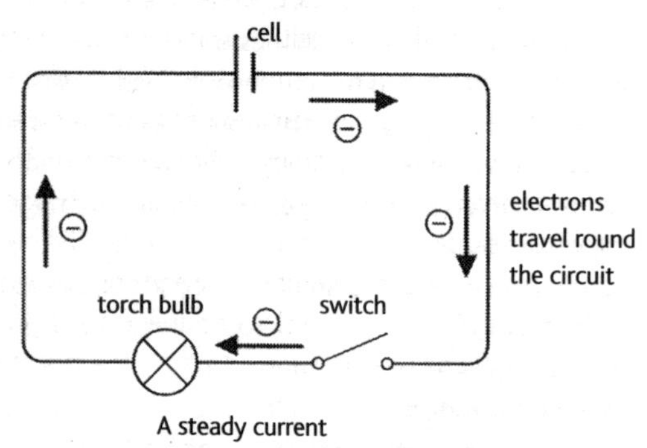

Figure 4.2 *Electrons at work.*

When a circuit is switched on, electrons leave the negative terminal of the battery, move round the circuit and re-enter the battery at its positive terminal. The one-way flow of charge round an electric circuit was first deduced by the French scientist, Andre Ampère. Before Ampère's discovery, it was thought that positive charge flowed from the positive terminal of the battery and was cancelled out by negative charge flowing from the negative terminal. Ampère knew that an electric current along a wire caused a magnetic compass near the wire to deflect and he knew that this was because a magnetic field was created around the wire when an electric current passed along it. He also noticed that the compass deflected in the opposite direction if the battery was reversed in the circuit. He realized that this observation could only be explained if only one type of charge flows round a circuit. Reversing the battery caused the flow of charge in the circuit to reverse, which reversed the direction of the magnetic field round the wire. However, Ampère was not able to say whether the charge was positive charge from the positive terminal or negative charge from the negative terminal of the battery. To prevent confusion, scientists agreed that the direction of current in a circuit should be the direction of flow of positive charge. We know now that an electric current in a wire is a flow of electrons round the circuit from the negative terminal to the positive terminal of the battery. Nevertheless, the rule that the direction of current is the direction of flow of positive charge is still used today.

The **unit of electric current** is the **ampere** (A), defined in terms of its magnetic effect, and can be measured using an ammeter. A current of 1 A passing along a wire is due to 6.25 million million million electrons passing along the wire every second. Each electron carries the same amount of negative charge. The bigger the current in a wire, the greater the number of electrons passing along the wire each second. Therefore, the charge passing along a wire in a certain length of time is proportional to the current and the duration of the time interval.

Note: Small currents are measured in *milliamperes* (mA) or *microamperes* (μA), where 1 mA = 0.001 A and 1 μA = 0.001 mA.

Batteries and cells

In any circuit, electrons transfer energy from the battery to the components in the circuit. Each electron in an electric circuit gains energy as it passes through the battery, leaves the battery via the negative terminal, loses energy as it passes round the circuit and re-enters the battery via its positive terminal. Each electron leaves the negative terminal of the battery with electrical potential energy, which is used up and converted to other forms of energy as the electron passes round the circuit. For example, when an electron passes through a torch bulb, energy is transferred from the electron to the torch bulb. The effect of all the electrons passing through the torch bulb filament is to make it so hot that it emits light. The connecting wires are good conductors so no energy is given up in these wires, provided the current is not excessive.

A **battery** consists of two or more identical **cells**, each cell consisting of two electrodes in a conducting paste or liquid, the electrolyte, which reacts with the electrodes. The two electrodes are made from different materials such as graphite and lead, chosen because they react differently with the electrolyte. One electrode loses electrons to the electrolyte and the other one gains electrons from the electrolyte. Thus one electrode becomes positively charged and the other one becomes negatively charged. In use, a cell gradually deteriorates because the chemical reactions between the electrodes and the electrolyte convert the electrolyte and the electrodes into other substances which do not react. In a rechargeable cell, this chemical process can be reversed by connecting the cell to a battery charger, which forces electrons into the cell at the negative electrode and out at the positive electrode. Disposable batteries and cells cannot be recharged and are discarded when exhausted.

The **voltage** of a battery or a cell is the power in watts that the cell can deliver to the rest of the circuit for every ampere of current passing through it. Thus a 12-volt battery connected in a circuit delivers 12 watts of power for every ampere of current passing through it. If it is connected to a suitable 3A light bulb, the battery supplies 36 watts of power to the light bulb because the current is 3 amperes.

Voltage (in volts) = power delivered (in watts) per ampere of current:

$$\text{Voltage or potential difference (in volts)} = \frac{\text{power delivered in watts}}{\text{current in amperes}}$$

The capacity of a battery or cell is usually expressed in ampere hours. This is the number of hours for which it would be able to supply 1 ampere of current before it is exhausted. Thus a car battery with a capacity of 60 ampere hours would be able to supply a current of 1 ampere for 60 hours. Such a battery could supply a current of 2 amperes for 30 hours or a current of 5 amperes for 12 hours.

Electric circuits

Circuit rules

A **series circuit** is one in which the same current passes through all the components in the circuit, as shown in Figure 4.3(a). In other words, all the electrons moving round such a circuit pass through every component in the circuit. The current may be measured using an **ammeter**, which is a meter designed to measure current. The ammeter must be connected in series with the other components in the circuit so that the same current passes through all the components in the circuit.

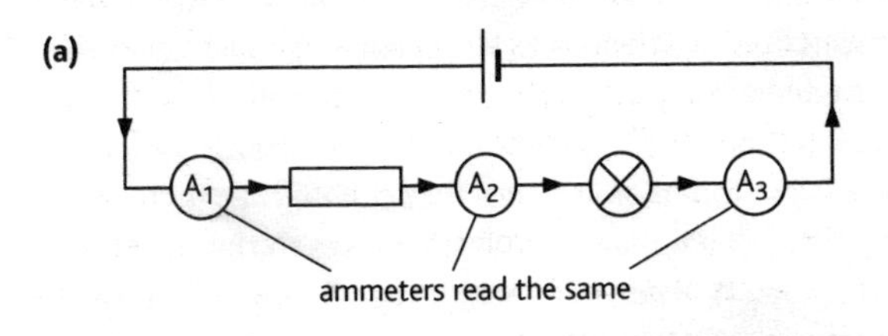

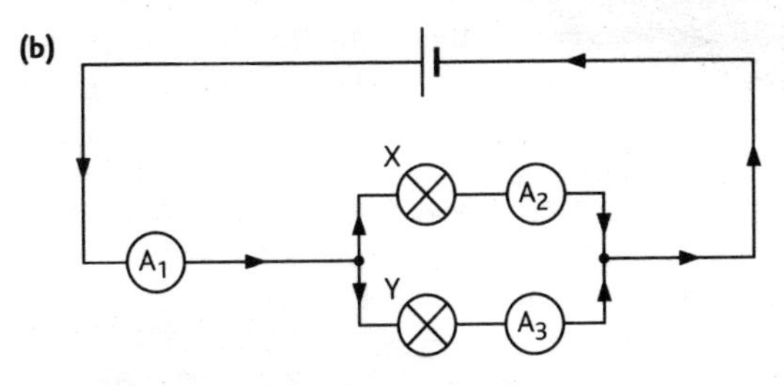

Figure 4.3 *(a) Components in series. (b) Components in parallel.*

A circuit with components **in parallel** with each other is one in which part of the current from the battery passes through each of the parallel components. See Figure 4.3(b). Components in parallel in a circuit have the same voltage between the two points where they are connected to the circuit. Each electron moving round the circuit passes through one component or the other component when it moves from one point to the other. The voltage between two points in a circuit may be measured using a **voltmeter**, which is a meter designed to measure voltage. To measure the voltage across a component, the voltmeter must be connected in parallel with the component.

* For components in series, the **current** is the same.
* For components in parallel, the **voltage** is the same.

Resistance

A potential difference or voltage exists between two points in an electric circuit if an electron has to do work to move from one point to the other. Conductors in a circuit resist the flow of electrons because the electrons repeatedly collide with the atoms of the conductor and lose energy in these collisions. The atoms of the conductor vibrate and the vibrations increase when the atoms gain energy from the electrons that collide with them. As a result, the conductor gains energy, which it loses to the surroundings through heat transfer. In effect, the conductor resists the flow of electrons through it and the electrons must use some of their energy to pass through the conductor. Thus if a voltage exists between two points in any circuit, the electrons must use some or all of their energy to overcome the resistance of that part of the circuit to their passage from one point to the other.

The **resistance** of a circuit component is defined as

$$\frac{\text{the voltage across the component}}{\text{the current through the component}}$$

The unit of resistance is the **ohm** (symbol Ω), which is the amount of resistance between two points in a circuit when the voltage between the two points is 1 volt and the current is 1 ampere.

$$\text{Resistance (in ohms)} = \frac{\text{voltage (in volts)}}{\text{current (in amperes)}}$$

Circuit diagrams

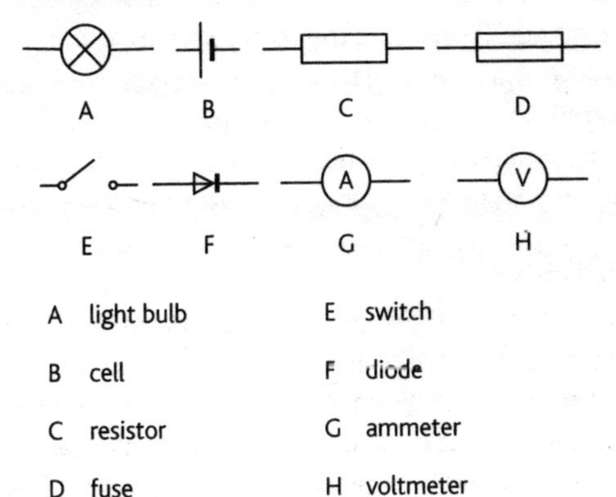

Figure 4.4 *Electrical symbols.*

Standard symbols are used to represent components in circuit diagrams. The function of each component represented in Figure 4.4 is described below, if it has not been previously described.

* A **resistor** is a component designed to have a known resistance.
* A **diode** allows current through in one direction only. A diode conducts when it is connected into the circuit in its forward direction, as its resistance in the forward direction is very low. If the diode is then reversed, it does not conduct because its resistance in the reverse direction is very high.
* A **light-emitting diode** (LED) emits light when current passes through it. LEDs are used as indicators in electronic circuits.

The electrical power equation

Because the voltage between any two points in a circuit is the power delivered per ampere, it follows that the power delivered for a certain current can be calculated by multiplying the voltage by the current. In other words,

Power delivered (in watts) = voltage (in volts) × current (in amperes).

For example, an electric kettle that operates at a voltage of 230 volts and a current of 10 amperes would take 2300 watts of power (= 230 volts × 10 amperes) from the electricity supply when it is switched on. A car's electrical heater designed to operate at 12 volts and 5 amperes would take 60 watts of power (= 12 volts × 5 amperes) from the car battery, if the engine was not switched on.

Electricity at work

A magnetic compass near a wire is deflected when the current in the wire is switched on. This is because the current creates a magnetic field round the wire. With no current in the wire, the compass needle points north because of the Earth's magnetic field. If the compass was moved in the direction of its needle points, it would follow a straight line northwards. Such a line is called a magnetic field line, or sometimes a magnetic line of force. Its direction is defined as the direction in which a compass needle points. With the current on, the compass would need to move round the wire because the magnetic field lines are circles centred on the wire, as shown in Figure 4.5. If the current is reversed in the wire, the direction of the magnetic field lines round the wire reverses.

Electromagnets

An electromagnet consists of a coil of insulated wire wound round an iron bar. When a current is passed through the coil, the

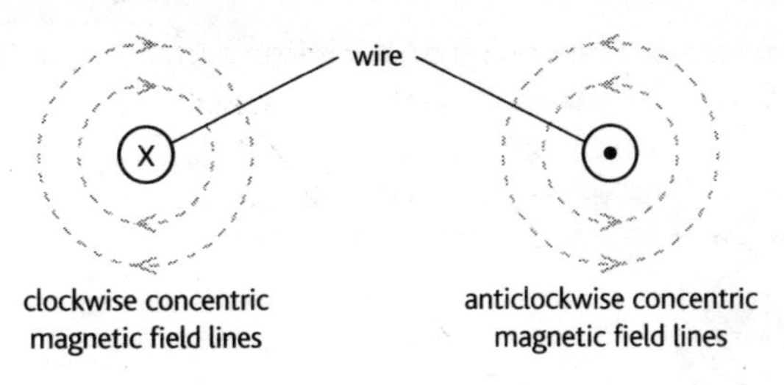

Figure 4.5 *The magnetic field around a current-carrying wire.*

iron bar is magnetized by the magnetic field due to the current. As a result, the iron bar is able to attract iron and steel objects held near its ends. When the current is switched off, the iron bar loses its magnetism. Electromagnets have many uses, from powerful electromagnets used to pick up and move cars in scrapyards to very sensitive electromagnets used to write data onto computer discs.

Electric motors

A wire near a magnet experiences a force when a current is passed along the wire. This effect is known as the **motor effect**. The reason for the effect is that the electrons passing along the wire are pushed sideways by the magnet, exerting a sideways force on the wire as a result. The effect depends on the angle between the wire and the lines of the magnetic field. For maximum effect, the wire needs to be at right angles to the lines of the magnetic field.

An electric motor consists of a coil of insulated wire, which spins between opposite poles of a U-shaped magnet or electromagnet. When current is passed round the coil, the coil turns because the wires along one edge of the coil are forced up and the wires along the opposite edge are forced down. After half a turn, the current round the coil must be reversed, otherwise the coil is turned back by the magnetic field. This reversal is achieved automatically by the action of the split-ring commutator which rotates with the coil. Conducting brushes made of graphite connected to the battery press against the commutator. Graphite is used as it is a conductor and provides contact with little friction.

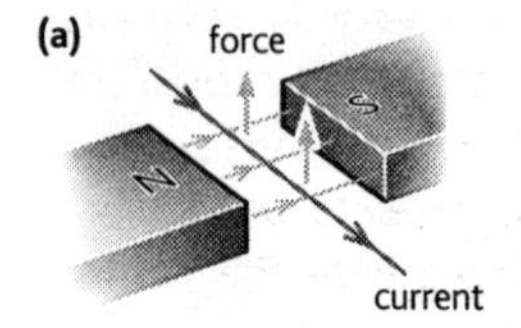

Current passes along each side of the coil in opposite directions. Each side is therefore acted on by a force due to the magnetic field. The force on one side is in the opposite direction to the force on the other side.

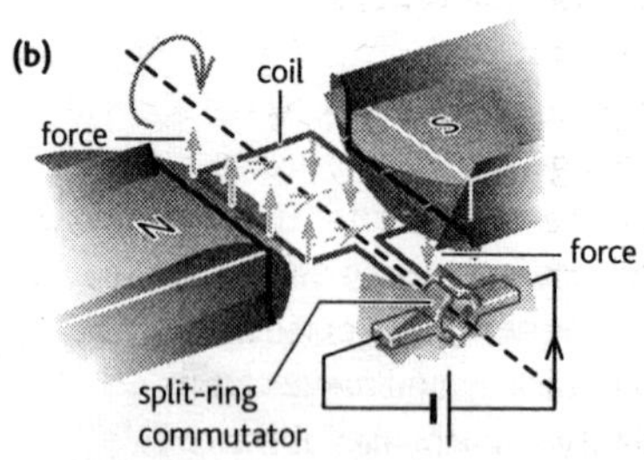

When the coil is parallel to the field, the forces on the sides rotate the coil. As the coil turns through the position at 90° to the field, the split-ring commutator reverses its connections to the battery, reversing the current direction round the coil.

Rotation by more than 90° has swapped the positions of the two sides and reversed the current direction in the coil. Therefore, the forces acting on each side continue to turn the coil in the same direction as before, so the coil rotates continuously in one direction.

Figure 4.6 *The motor effect and the electric motor.*

The split-ring commutator has two functions:

1 It provides continuous contact between the battery and the coil as the coil turns.

2 It reverses the current direction round the coil every half turn, so the coil is forced to spin in the same direction every half turn. Without this reversal of the current direction every half turn, the coil would reverse its direction repeatedly.

If the battery is reversed in the circuit, the motor spins in the opposite direction. The greater the current, the faster the rate of rotation of the coil. In a mains electric motor:

* an electromagnet is used instead of a permanent magnet
* a number of coils at equal angles are wound on an iron core, thus providing smoother rotation. Each coil is connected to opposite sections of a multi-segment commutator.

Generators

Electricity can be generated in a coil by moving the coil in a magnetic field. Provided the coil wires cut the lines of the magnetic field, a voltage is induced in the coil. If the coil is part of a complete circuit, the induced voltage makes a current pass round the circuit. A cycle **dynamo** consists of a magnet that is forced to rotate near the end of a coil, causing a voltage to be induced in the coil. The faster the magnet rotates, the greater the voltage.

The **alternating current generator** consists of a rectangular coil of insulated wire that is forced to rotate between the poles of a U-shaped magnet. The wires along opposite edges of the coil cut across the magnetic field lines as the coil rotates, causing a voltage to be induced in the coil. The coil wires are connected to two slip rings on the axle of the coil. A graphite 'brush' pressed against each slip ring provides a continuous contact to an external circuit.

The voltage reverses in polarity as each edge of the coil crosses the middle from one magnetic pole to the other pole every half turn. The voltage peaks when the coil edges are nearest the poles. Thus the voltage varies as shown, acting in each direction every half turn. This type of voltage is referred to as an *alternating voltage*. The voltage is said to 'alternate' because it repeatedly reverses its direction. If the generator is connected to a suitable resistor, an alternating current is forced round the circuit, reversing its direction repeatedly. For each complete turn of the coil, the current reverses and reverses back again in a full cycle.

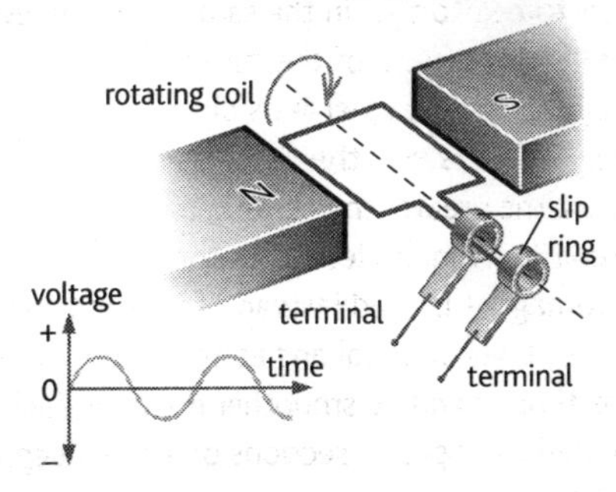

Figure 4.7 *The a.c. generator.*

5

the nature of light

In this chapter, we will start by studying reflection and refraction of light and the formation of images. We'll then move on to consider two different theories put forward over 300 years ago about the nature of light, both of which can explain reflection and refraction. One theory said that light was made up of tiny particles. The other theory said that light consists of waves. Which theory proved to be correct? We will look at the evidence for and against each theory and how, eventually, a century later, more evidence was obtained that led to the acceptance of one theory and rejection of the other.

The story of the nature of light had further twists in store with the discovery of photoelectricity at the end of the nineteenth century. Investigations into this effect produced observations that could not be explained until a new theory of light, namely the photon theory, was proposed by Albert Einstein.

Properties of light

A laser beam or a beam of sunlight travels in a straight line. On a cloudy day, sunlight that breaks through a gap in the clouds can be seen as a straight beam extending to the ground. We talk about the Sun's rays, but what do we mean by the word 'ray'? It seems a very convenient word for the path that light takes when it radiates from a source (e.g. light from the Sun). So for convenience we will think of light in terms of rays. Also, let us assume for now that light travels in straight lines.

Reflection of light

Stand in front of a flat mirror and you will see an image of yourself. You will also be able to see beyond your own image the image of any other object behind you. Each image is the same distance behind the mirror as the object is in front. So if you stand 50 centimetres in front of a flat mirror, your image is 50 centimetres behind the mirror and is therefore 100 centimetres away from you. Step back from the mirror and your image steps back by the same distance.

Mirror images

Make a semi-transparent 'mirror' from cellophane film over an open box. Put the box on its side so the film is vertical and place a small object in front of it. You should be able to see an image of the object in the box. Move the object about until its image appears to be at the back of the box. You should find that the distance from the object to the 'mirror' is then the same as the distance from the 'mirror' to the back of the box.

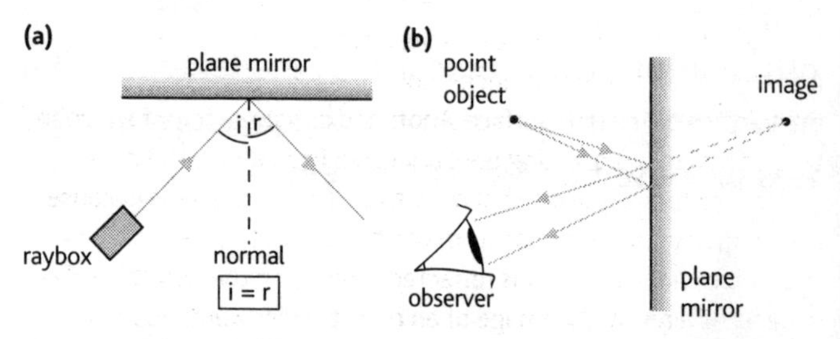

Figure 5.1 *(a) The law of reflection. (b) Image formation by a plane mirror.*

When a ray of light is directed at a certain angle onto a flat mirror, the light ray reflects off the mirror at the same angle. In other words, the angle between the mirror and the light ray before reflection is the same as the angle between the mirror and the light ray after reflection. This statement is known as the **law of reflection** and it holds for any mirror. The law is usually expressed in the form

$$i = r$$

where i is the angle between the incident ray and the **normal** (the line at right angles to the mirror at the point of incidence) and r is the angle between the reflected ray and the normal.

Using this law, we can explain the distance rule by means of a ray diagram as shown in Figure 5.1(b). This diagram represents two light rays from a point object before and after reflection. The angle between each light ray and the mirror is the same before reflection as after reflection. Someone looking into the mirror along both reflected rays would see an image of the object at the position where the reflected rays appear to come from. The mirror in the ray diagram in Figure 5.1(b) is a line of symmetry. This is because each ray from the object to the mirror is at the same angle to the mirror as the corresponding 'apparent ray' from the image to the mirror. Therefore, the image and the object are the same distance from the mirror.

Refraction of light

Refraction of light is the change of direction of a light ray when it passes from one transparent substance to another transparent substance. For example, when light passes from air into a lens, its direction changes, unless its initial direction was at right angles to the lens surface. Another example can be seen when the bottom of a swimming pool is viewed from above the water surface. The pool appears shallower than it really is. This is because light from a point on the bottom of the pool that passes through the surface non-normally is refracted away from the normal at the surface. As a result, the image of an object under water appears closer to the surface than the object really is.

Light is refracted

* towards the normal when it passes from air into a transparent substance
* away from the normal when it passes from a transparent substance into air.

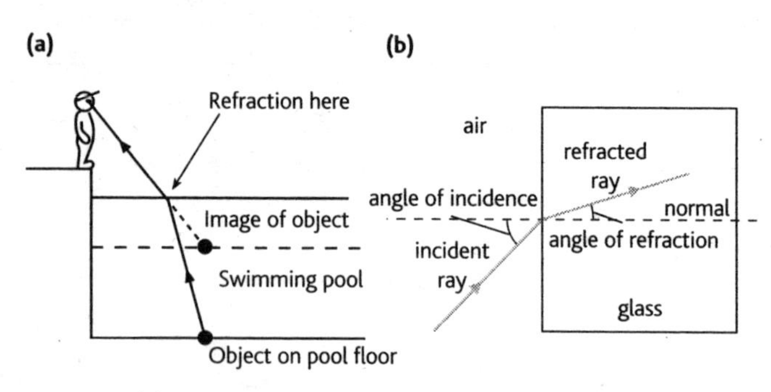

Figure 5.2 *(a) Refraction at a water surface. (b) Testing the law of refraction.*

Theories of light

Newton's theory of light

If you have played or observed a game of snooker, you will know that a ball bounces off a wall at the same angle as it hits the wall – provided that it is not spinning as it rolls towards the wall. In addition to establishing laws on motion and gravity, Sir Isaac Newton put forward a theory of light in which he proposed that light consists of tiny particles that he called **corpuscles**. Perhaps Newton had the idea in his mind about a ball bouncing off a wall, for that was how he explained the reflection of light, supposing that the corpuscles of light bounced off a mirror in the way that a ball bounces off a wall. Newton needed to assume that the corpuscles did not lose any speed due to the impact so they moved away from the wall at the same angle as they had moved towards it.

Newton used his corpuscular theory to explain refraction of light as well as reflection. According to Newton, a corpuscle of light in air moving towards a transparent substance is attracted towards the substance. The force of attraction causes it to move faster in the substance than in air. More importantly, if its initial direction is not directly along the normal, the increase of speed causes its direction to change nearer to the normal.

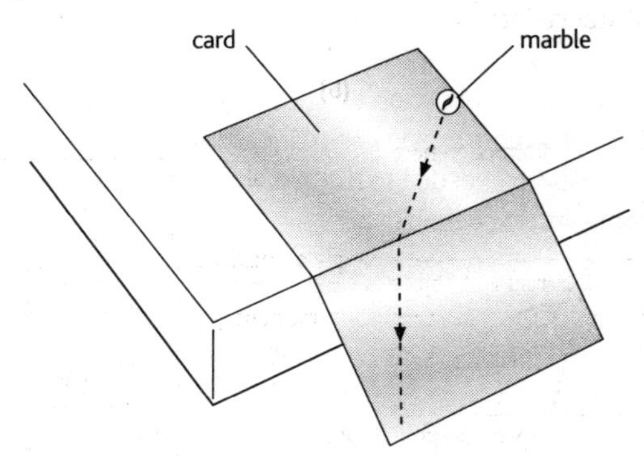

Figure 5.3 *A refraction model.*

A model of refraction can be made using a marble and a piece of card folded down the middle. One half of the card needs to be fixed to a book on a slight slope, with the other half of the card sloping steeply down to the surface on which the book rests. Observe the progress of a marble rolled slowly across the card on the book towards the fold. When it passes over the fold onto the steeper section, its speed increases and its direction changes at the fold.

Huygens' wave theory of light

About the same time as Newton put forward the corpuscular theory of light, an alternative theory of light was put forward in Holland by Christiaan Huygens. According to Huygens, light consists of a wave motion that moves through space or air or any transparent substance in the way that waves move across a water surface. This imaginative theory also provided an explanation of the laws of reflection and refraction of light. Reflection of water waves is observed when sea waves bounce off a harbour wall. Refraction of water waves is the reason why waves on a beach usually run straight up the beach, regardless of the direction of the waves as they approach the beach. The waves slow down as the water becomes less deep and this reduction in speed causes them to move directly towards the beach. This situation is not unlike a front-wheel drive vehicle that veers off a road onto muddy ground. The vehicle changes direction as it goes off the road because one of the front wheels leaves the road and loses some of its grip before the other one does. To explain refraction of light, Huygens needed to assume that light travels more slowly in a transparent substance than in air. In contrast, Newton needed to assume that light travels faster in a transparent substance than in air.

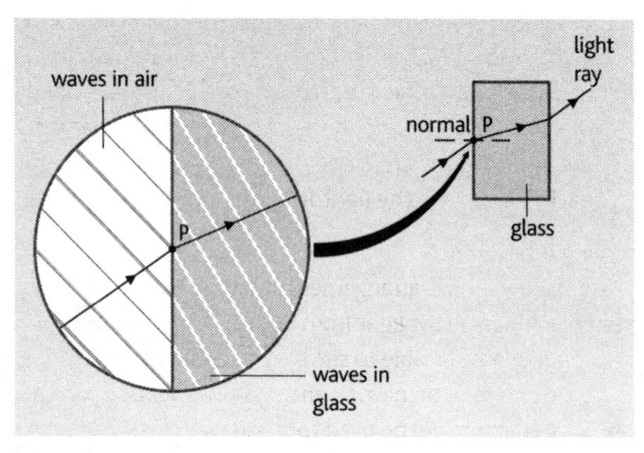

Figure 5.4 *Using wave theory to explain refraction.*

Young's double-slits experiment

Reflection and refraction of light can both be explained using either Newton's corpuscular theory or Huygens' wave theory of light. Newton's theory assumes that light travels faster in a transparent substance than in air, whereas Huygens' theory assumes that it travels more slowly in a transparent substance. Which theory is correct? When the theories were compared in the seventeenth century, most people believed Newton, as the speed of light could not be measured and Newton's scientific reputation was much greater than that of Huygens. Newton's theories of motion and gravity were outstandingly successful in explaining the motion of objects in all possible situations. So Newton's theory of light was accepted for over a century, even though there was no direct evidence for the corpuscular theory of light.

The idea that light consists of tiny particles held sway until the first decade of the nineteenth century, when Thomas Young at the Royal Institution in London used light to demonstrate the phenomenon of **interference**. Young showed that if a narrow source of light is observed through two closely spaced slits, a series of bright and dark bands (referred to as 'fringes') are seen. The fringes are replaced by a broad band of light if one of the two slits is blocked. Thus the dark bands are formed where light from one slit cancels out light from the other slit. The light from the two slits is said to interfere

* constructively, where a bright fringe is formed – the light from each slit reinforces the light from the other slit
* destructively, where a dark fringe is formed – the light from each slit cancels the light from the other slit.

Observing interference

Use the double slit arrangement shown in Figure 5.5 to observe interference fringes using light from a torch lamp adapted as a narrow source. You ought to be able to see bright and dark parallel fringes.

The slits need to be narrow and less than about 0.5 mm apart. A fine-pointed pin could be used to make two closely spaced slits in a strip of metal foil. The light source needs to be as narrow as possible, e.g. a thin gap between insulation tape over the front of a torchlamp could be used.

The phenomenon of interference can also be demonstrated by sending straight waves on a water surface towards two narrow closely spaced gaps in a barrier in the water. The waves that pass through the gaps spread out and overlap. In the overlap region, wave crests and troughs from one slit overlap and pass through crests and troughs from the other slit.

* Reinforcement occurs where a crest meets a crest or a trough meets a trough. With light, this occurs at each bright fringe.
* Cancellation occurs where a crest meets a trough. With light, this occurs at each dark fringe.

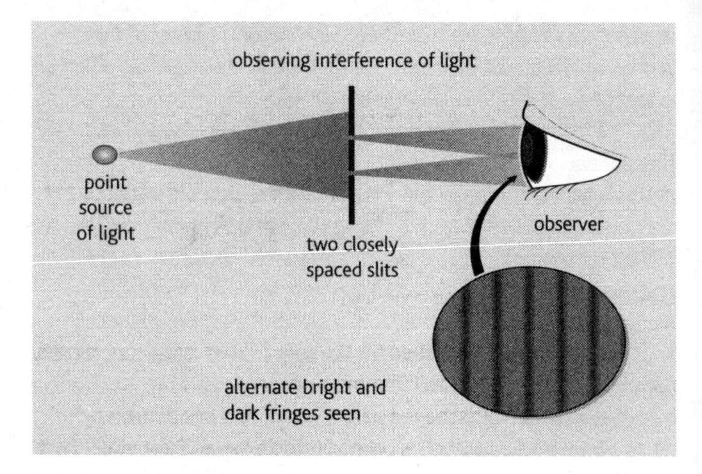

Figure 5.5 *Observation of interference fringes.*

Electromagnetic waves

Water waves are disturbances that travel across the surface. Sound waves in air are pressure variations that travel through the air.

Seismic waves are vibrations that travel through the Earth after being created in an earthquake. What are light waves? In the nineteenth century, it was generally thought that light consists of vibrations in an invisible substance referred to as 'ether', which was thought to fill space. In 1862, a mathematical theory of light waves was published by the Scottish physicist James Maxwell. He showed that light consists of vibrating electric and magnetic fields in which the electric vibrations generate magnetic vibrations which generate electric vibrations, and so on. Such waves were referred to as **electromagnetic waves**. Maxwell combined the theory of electric fields with the theory of magnetic fields to show that such electromagnetic waves ought to travel through space at a speed of 300 000 kilometres per second. He knew that light travels at this speed through space, so he concluded that

* light consists of electromagnetic waves, and
* electromagnetic waves exist beyond both ends of the visible spectrum.

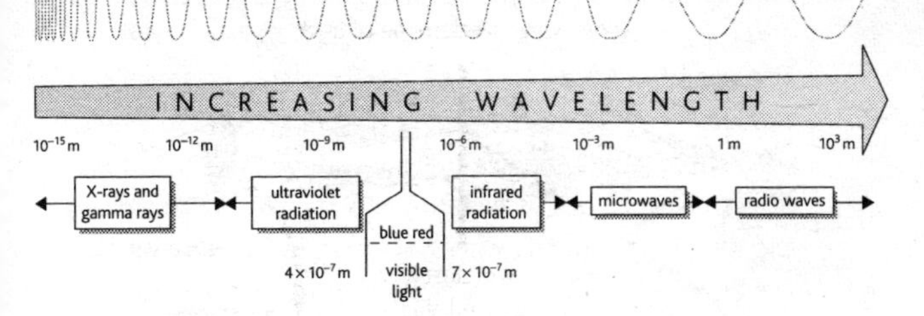

Figure 5.6 *The electromagnetic spectrum.*

When Maxwell published his theory of electromagnetic waves, it was immediately realized that infrared radiation, which had been detected beyond the red part of the visible spectrum, and ultraviolet radiation, which was known to lie beyond the violet part of the visible spectrum, must also consist of electromagnetic waves. Could there be electromagnetic waves even longer in wavelength than infrared radiation? What about beyond ultraviolet radiation? Several decades after Maxwell published his theory, the German physicist Heinrich Hertz discovered how to produce and detect electromagnetic waves much longer in wavelength than infrared radiation. These waves became known as radio waves and within a few years they were being used to transmit wireless signals between Britain and the USA.

Tuning in

When you tune in to a radio or TV station, you are adjusting a receiver circuit so that it will accept radio or TV waves of a certain wavelength only. The **wavelength** of a wave is the distance along the wave from one wave crest to the next wave crest. The Greek symbol λ (pronounced 'lambda') is used for wavelength. The tuner of a radio receiver usually displays the wavelength or the frequency of the waves that the tuner accepts.

* The **frequency** is the number of complete cycles of waves passing a point each second, where one cycle is from one wave crest to the next wave crest. The unit of frequency is the hertz (abbreviated as Hz), where 1 Hz = 1 cycle per second.

Local radio stations broadcast at frequencies of about 100 megahertz (MHz), where 1 MHz = 1 million hertz.

* The **speed** of waves is the distance per second travelled by a wave crest. For waves moving at speed v, in 1 second a wave crest would travel a distance equal to v. The number of wavelengths in this distance is equal to v/λ, where λ is the wavelength of the waves. Hence the frequency f of the waves, which is the number of waves passing a given point in 1 second, is equal to v/λ.

Frequency, f = speed of waves, v/wavelength, λ

Hence for electromagnetic waves of wavelength λ, the frequency of the waves $f = c/\lambda$, where c is the speed of electromagnetic waves. For transmission through air or through space, c = 300 000 kilometres per second.

X-rays

A few years after the discovery of radio waves, the German physicist Wilhelm Röntgen discovered how to produce and detect electromagnetic waves much shorter than ultraviolet radiation. These waves became known as **X-rays**. Their discovery received great publicity as newspapers claimed they could be used to 'see' through objects. In fact, X-ray machines were quickly put to use in hospitals to photograph broken bones in limbs. This is possible because X-rays are absorbed by dense materials such as bone and pass through soft material such as human tissue. If a beam of X-rays is directed at photographic film in a light-proof wrapper, the film is blackened where X-rays reach it, because the X-rays, unlike light, can pass through the wrapper. If a limb is placed in the path of the beam before it reaches the film, a shadow of the bone in the limb can be seen on the film when it is developed, because the bone prevents X-rays from reaching the film.

In the same decade as X-rays were discovered, the discovery of radioactivity by Henri Becquerel in France led to the conclusion that radioactive substances can emit electromagnetic waves even shorter than X-rays. This type of radiation is known as **gamma radiation** (or γ radiation using the Greek letter γ, pronounced 'gamma'). Maxwell's theory of electromagnetic waves was thus confirmed by these discoveries. Predictions from the theory were used to develop many practical applications, such as the design of radio transmitters on ships and of X-ray tubes for use in hospitals.

Classical physics based on Newton's laws and Maxwell's theory of electromagnetic waves was able to account for every known phenomenon or observation. Many physicists towards the end of the nineteenth century thought that the laws of nature had more or less been discovered, using the theories of classical physics. It seemed that little remained to be explained, although more accurate measurements of the properties of materials and light would probably justify the scientists' continued efforts. Some niggling minor discoveries were proving troublesome to explain, but by the turn of the century many physicists were satisfied that the laws of nature were mostly known.

Photoelectricity and photons

When Hertz was investigating how to produce and detect radio waves, he noticed that the sparks induced by radio waves in a 'spark-gap' detector were stronger when ultraviolet radiation was directed at the spark gap contacts. Hertz was more interested in radio waves and so he passed his observation on to other physicists to investigate. Further investigations showed that the effect happens because a metal emits tiny negative particles identified as electrons when illuminated with light of frequency above a certain value. This happens when ultraviolet radiation is directed at a metal because ultraviolet has a much higher frequency than any colour of light. The investigators knew that electrons are contained in every atom. They also knew that metals conduct electricity because some of the electrons in a metal move about inside the metal, not confined to individual atoms. Now they had found that light could be used to make some of these electrons escape from the metal. This effect is called the **photoelectric effect**.

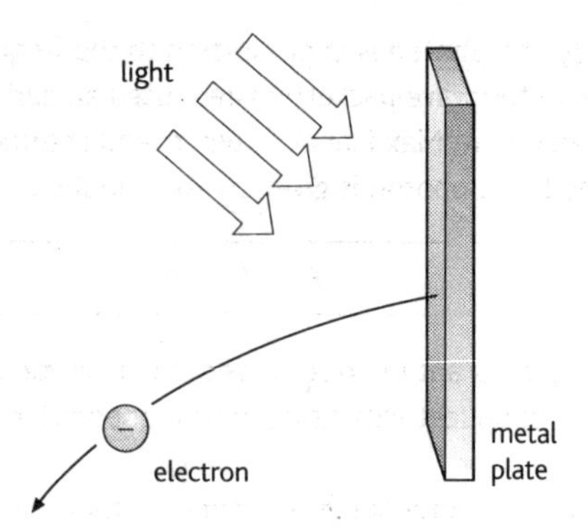

Figure 5.7 *Photoelectricity.*

Precise observations using light of different frequencies produced the astonishing conclusion that the frequency of the light needed to be greater than or equal to a certain 'threshold' value that depended on the metal being tested. The existence of such a threshold frequency could not be explained using the wave theory of light. According to wave theory, the effect should happen regardless of the frequency of the light. It was expected that the effect would happen more slowly, the lower the frequency of light used – but nevertheless it would happen, according to wave theory. However, the experimental results showed that electrons were not emitted at all if the frequency of the light was below the threshold frequency. In 1905, this troublesome problem from the last few years of the nineteenth century was to shatter the well-established classical wave theory of light.

Einstein's photon theory of light

Einstein explained photoelectricity in his 1905 paper on the interaction of light and matter by inventing a new theory of light which he called the **photon theory of light**. Einstein assumed that light is composed of wave packets that he called photons. The key points of photon theory are:

1. Each photon is a packet of electromagnetic waves moving in a particular direction, not spreading in all directions as in classical wave theory.

2. The energy of a photon is in proportion to the frequency of the waves in the wave packet. Einstein used an earlier idea about energy from Max Planck in Berlin, and assumed that the energy E of a photon is given by the equation

$$E = hf$$

where h is a constant referred to as the Planck constant.

To explain photoelectricity using photon theory, Einstein said that:

1. An electron in a metal needs a minimum amount of energy to escape. This amount of energy is called the work function of the metal.
2. An electron near the surface can escape if it gains energy equal to or greater than the work function of the metal.
3. An electron is able to escape from the surface if it absorbs a single photon of energy greater than the work function of the metal.

Thus if an electron near the surface absorbs a photon of energy hf, it can escape if hf is greater than or equal to W, the work function of the metal. The threshold frequency of the light therefore is equal to W/h, corresponding to $hf = W$.

Hence, for photoelectric emission to occur, the frequency of the incident light f must be greater than or equal to the threshold frequency. Thus Einstein's photon theory provides an explanation of photoelectricity. Further predictions about the interaction of light and matter were confirmed by more investigations after 1905, and by 1921 the photon theory had been fully accepted.

6

materials and molecules

Many new materials have been discovered or invented since 1900 by scientists extending their knowledge about existing materials. Before the Scientific Age, most materials were natural products or made from minerals. The Scientific Age brought new processes, new products and new discoveries, which have been put to use to make our lives better. A list of materials used by each of us every day would be very long. So it's all the more surprising that all materials and substances, manufactured or natural, are made from no more than 92 basic substances known as chemical elements. Progress to our present knowledge is due to the efforts of many scientists whose experiments and observations gave rise to scientific theories to explain the experimental data and then to further predictions and more experiments. New materials continue to be discovered in the same way, often leading to new devices and products.

Atoms and molecules

How would you describe the differences between solids, liquids and gases? Solids have their own shape whereas liquids and gases do not. Liquids and gases can flow whereas solids cannot. Liquids and solids have surfaces whereas gases do not. The three states of matter, solid, liquid and gas, are described as physical states because a substance can be changed from one state into another and back into its former state by heating and cooling. For example, ice melts and becomes water if heated. In contrast, chemical changes involve reactions between substances that cannot be reversed by simple physical means such as heating and cooling. For example, water and carbon dioxide are formed when a candle burns, but candle wax cannot easily be formed from water and carbon dioxide. The obvious differences between the three states of matter are explained using two key ideas:

* All substances consist of **molecules**. A molecule is the smallest particle of a substance that can exist independently. The molecules of a pure substance are identical to one another and differ from the molecules of any other substance.
* Molecules link together in the liquid state and the solid state because they exert forces on each other. These forces are referred to as **bonds**, because they bind the substance in the liquid or the solid state. Energy supplied to a solid at its melting point causes it to melt because the molecules gain sufficient energy to break free from each other. Energy supplied to a liquid at its boiling point causes it to vaporize because the molecules gain sufficient energy to move away from each other.

Compounds and elements

A **compound** is a pure substance that consists of one type of molecule only. A compound can be broken down into other substances.

An **element** is a pure substance that cannot be broken down into other substances. There are just 92 naturally occurring elements. Further short-lived elements have been discovered in nuclear reactions, but none of these is found naturally.

An **atom** is the smallest part of an element that is characteristic of the element. The atoms of an element are identical to one another and differ from the atoms of any other element. The lightest atom is the hydrogen atom. The heaviest naturally occurring atom is the uranium atom, which is 238 times heavier than the hydrogen atom.

A **molecule** consists of two or more atoms joined together by force bonds. Each type of molecule consists of a fixed number and type of atoms. For example, every carbon dioxide molecule consists of one carbon atom and two oxygen atoms. Its chemical formula is therefore written as CO_2.

Inside an atom

By the end of the nineteenth century, scientists knew that atoms are not indivisible or indestructible. Experiments on electrical discharges through gases at very low pressures resulted in the discovery of electrons. They were pulled out of the gas atoms by the strong electrical field applied to the discharge tube. Regardless of the type of gas used, it was discovered that identical negatively charged particles were produced as well as positively charged atoms of differing weights. It was therefore concluded that these negatively charged particles, referred to as electrons, are in every type of atom. Further investigations on radioactivity (see Chapter 7) led to the conclusions that:

* Every atom contains a positively charged nucleus where most of its mass is concentrated.
* The nucleus is composed of two types of particle, protons and neutrons. The nucleus of the hydrogen atom is a single **proton**.
* The proton is a positively charged particle and is slightly lighter than the **neutron**, which is uncharged.

* Electrons move in the space round the nucleus at relatively large distances. The charge of the electron is equal and opposite to the charge of the proton. The electron is about 2000 times lighter than the proton.

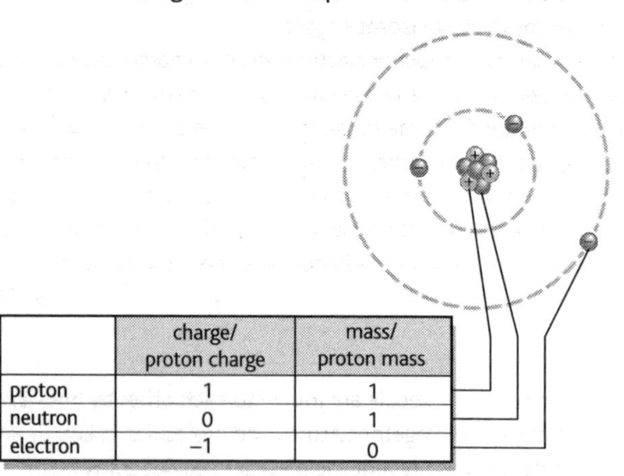

	charge/ proton charge	mass/ proton mass
proton	1	1
neutron	0	1
electron	–1	0

Figure 6.1 *The structure of the atom.*

An atom is represented by the symbol $^{A}_{Z}X$ where Z is the proton number of the atom and A, the mass number, is the number of protons and neutrons in the nucleus. The mass of an atom is approximately equal to A times the mass of a hydrogen atom. This is because A is the number of protons and neutrons in its nucleus and each proton or neutron has a mass approximately equal to the mass of a hydrogen atom. An uncharged atom has the same number of electrons and protons. For example, the symbol $^{7}_{3}\text{Li}$ represents an atom of lithium with three protons and four neutrons in its nucleus and three electrons moving about the nucleus. The mass of such an atom is approximately seven times the mass of a hydrogen atom.

Isotopes

Each atom of the same element contains the same number of protons in its nucleus. The number of neutrons in each atom of an element can differ. Atoms of an element with different numbers of neutrons are referred to as **isotopes**. For example, natural uranium mostly consists of the isotope $^{238}_{92}U$ and a small proportion of the isotope $^{235}_{92}U$. Both types of atoms are uranium atoms, each nucleus containing 92 protons. However, the isotope $^{238}_{92}U$ contains three more neutrons than the isotope $^{235}_{92}U$.

Note that the number of electrons in an uncharged atom is equal to the number of protons in the nucleus. The chemical properties of an element are the same for all the isotopes of the element. This is because chemical reactions are determined by the electrons in an atom. Atoms of the same element undergo the same chemical reactions, because each atom has the same electron arrangement even if the atoms are different isotopes of the same element.

More about bonds

The atoms in a molecule are joined to each other by **bonds**, which hold the atoms together. Atoms and molecules in liquids and solids form bonds that prevent the particles moving away from each other. If the bonds are sufficiently strong, the molecules lock each other together in a rigid structure. There are several different types of bond. The electron arrangement of each type of atom determines the type of bond formed and this determines the physical state of the substance at a given temperature.

The electrons in an atom move round the nucleus like planets moving round the Sun. The electrical attraction between each electron and the nucleus prevents the electron from leaving the atom. The electrons in an atom can only occupy certain orbits round the nucleus. These allowed orbits are called **shells**. Each shell can hold up to a certain number of electrons. The innermost shell of an atom can only hold two electrons. The next shell from the nucleus can hold up to eight electrons. The nearer a shell is to the nucleus, the lower the energy of an electron in the shell.

The chemical properties of an element depend on the electron arrangement of its atoms. Because a completely empty or full shell is in a lower energy state than a partially full shell, atoms react by gaining or losing electrons to make full shells only.

For example, lithium, sodium, potassium and certain heavier elements are metals that readily react because each type of atom in this group has a single electron in the outer shell. This lone electron is easily removed when the atom reacts with another atom. Another distinctive group of elements are the inert gases: helium, neon, argon, xenon and krypton. Atoms of all these elements have their electrons in completely filled shells and are therefore unreactive.

Types of bond

The type of bond formed between two atoms or molecules depends on how the atoms or molecules can lose or gain or share electrons to reach a lower energy state.

* A **covalent bond** is formed as a result of two atoms sharing a pair of electrons. Each atom contributes an electron to the bond from its outer shell. For example, the carbon dioxide molecule consists of one carbon atom and two oxygen atoms. Each oxygen atom forms two covalent bonds with the carbon atom. The second shell of the carbon atom in a molecule is therefore filled with eight electrons, four originally belonging to the carbon atom and two from each oxygen atom. The second shell of each oxygen atom in the molecule also has eight electrons, six from the oxygen atom and two from the carbon atom.

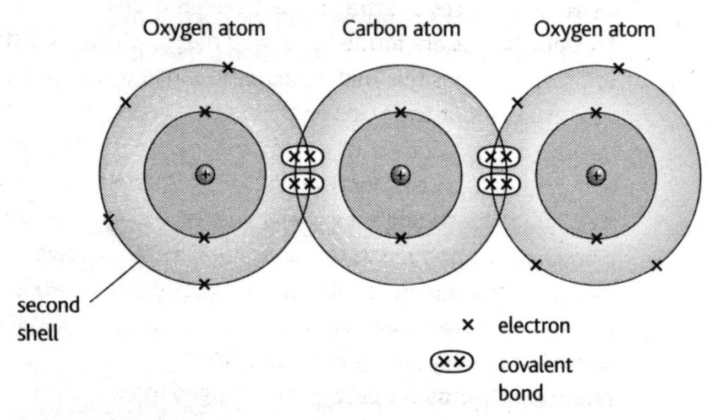

Figure 6.2 *Covalent bonds.*

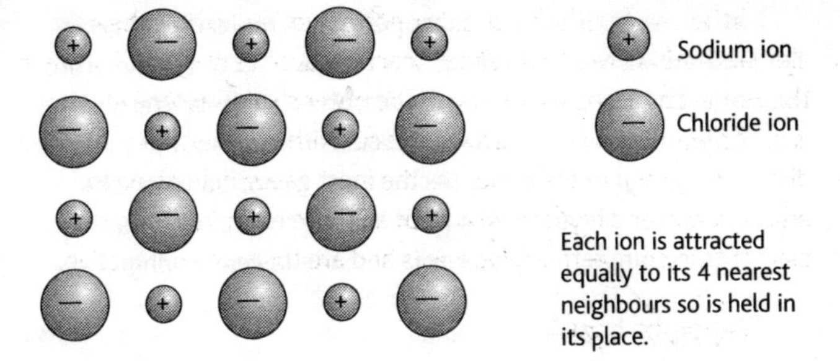

Figure 6.3 *Ionic bonding.*

* An **ionic bond** is formed when an atom of one element loses an electron in its outer shell to an atom of a different element. The donor atom becomes positively charged because it loses an electron, whereas the atom that gains the electron becomes negatively charged. Charged atoms are called **ions**. Hence the atoms that form ionic bonds attract each other because they are oppositely charged. The characteristic shape of a crystal occurs because it contains positive and negative ions arranged in a regular pattern referred to as a lattice. Each **ion** is held in place in the lattice by ionic bonds formed with adjacent, oppositely charged ions. Ionic crystals dissolve in water because water molecules weaken the forces of attraction between the ions.
* **Metallic bonds** are formed in a solid metal, which consists of a lattice of positive metal ions surrounded by electrons that move about freely inside the metal. Each metal atom has lost one or more of its outer electrons. These electrons are referred to as **conduction electrons** because they carry electric charge through the metal when a voltage is placed across it. The conduction electrons prevent the positive ions from moving out of place in the lattice. Solid metals are much stronger than many other solids because metallic bonds are equally strong in all directions.
* **Molecular bonds** act between uncharged molecules at close range. This type of bond is much weaker than the

other types of bond described above. A molecular bond is formed when two molecules are so close that the electrons of each molecule are slightly attracted to the nucleus of the other molecule. Molecules in liquids move about at random, but remain in the liquid because the bonds between the molecules prevent them from leaving the liquid surface. If the liquid temperature is raised, the faster-moving molecules near the surface can break away from the liquid and become gas molecules.

Solids and structure

A solid object has its own natural shape because the force bonds lock the atoms of the solid together. When a solid object is stretched or compressed or twisted, the atoms are pulled away from each other. If the forces distorting an object are removed and the object returns to its natural shape, the object is said to possess **elasticity**. If the distorting forces are sufficiently large, the object will not regain its natural shape and is permanently distorted. For example, a Perspex ruler that is bent slightly becomes straight again when the bending forces are removed. If the bending forces are large enough, the ruler bends permanently or cracks. Another example is a paper clip, which is designed to be strong enough to hold sheets of paper together but not too strong or it would be impossible to use. If too much force is applied to a paper clip, it bends permanently. The limit beyond which an object loses its elasticity is referred to as its **elastic limit**. Above its elastic limit, the object is permanently distorted and is said to be **plastic** rather than elastic. The bonds between atoms in the object break and are re-formed with different atoms when the object shows plastic behaviour.

Atoms in solids

Solids may be classified as crystalline, amorphous or polymers.

1 Crystalline solids

The atoms in a crystal are arranged in a regular pattern. As a result, the crystal has a recognizable shape. For example, sodium chloride crystals are cubic because sodium and chloride ions occupy alternate positions – like opposing sports fans next to each other in rows of seats. Imagine each 'reds' fan sitting with an opposing 'blues' fan immediately in front, behind and on either side. The analogy ends there, as the sodium ions are much smaller than the chloride ions and carry opposite charge, which is why the structure holds together. Crystals with a particular shape consist of rows of atoms arranged in such a way as to create the characteristic shape of the crystal.

2 Amorphous solids

An amorphous object has no characteristic shape. The atoms in an amorphous solid are locked together in contact with each other at random. In a liquid, the atoms are in contact with each other but they are not locked together and they move about in random directions. An amorphous solid is sometimes described as a 'frozen liquid', as if the atoms of a liquid suddenly stopped moving and locked together haphazardly. Glass is an example of an amorphous solid, consisting of silicon atoms and oxygen atoms linked together by covalent bonds. Glass is brittle because the effects of stress in the glass, due to bending forces, concentrate at tiny cracks in the surface. These cracks then deepen, which causes the stress in the glass to become even more concentrated so the glass snaps suddenly.

3 Polymers

A polymer consists of long molecules, each consisting of a chain of atoms, formed as a result of identical shorter molecules joining together end-on. For example, polythene consists of long polyethylene molecules formed by making ethylene molecules join end-on. Each ethylene molecule consists of two carbon atoms and four hydrogen atoms. The two carbon atoms are joined together by a double bond (i.e. two covalent bonds) and two hydrogen atoms are attached to each carbon atom. When the ethylene molecules join end-to-end, the double bond is replaced by a single bond, which enables the carbon atoms to link together in a long row. Each carbon atom in the polymer molecule is thus joined by a single covalent bond to a carbon atom on either side and by a covalent bond to each of two hydrogen atoms.

Molecules in fluids

Any substance that can flow is a fluid. The molecules in a fluid are not locked together and they move about at random. In a gas, the molecules move very fast and are separated by large empty spaces. The molecules in a liquid move about more slowly than in a gas and they are in contact with each other. Liquids (and solids) are much denser than gases because the atoms and molecules in liquids (and solids) are much closer to each other than they are in gases. The forces between gas molecules are too small to affect the fast-moving molecules in a gas. However, in liquids and solids, the molecules move more slowly than in a gas and the forces between them are strong enough to pull the molecules together.

Diffusion

Observe a few drops of milk spreading out in a glass of water without stirring the water. The milk gradually spreads throughout the water. This process of the spreading of a fluid in another fluid is known as **diffusion**. It occurs because the molecules of each fluid are in motion and so they move about until they are evenly distributed. Diffusion takes place with gases as well as with liquids. For example, vapour from a liquid with a strong odour spreads out through the surrounding air. Someone near the liquid would notice the odour before someone further away. Initially, the vapour molecules are concentrated near the surface of the liquid but gradually they move away from the liquid in every direction until they are evenly distributed throughout the air. This process is the same as what happens when a container is filled with a layer of red beads covered with a layer of blue beads. If the lid is then closed and the container is shaken repeatedly, the beads gradually move about until they are distributed at random throughout the container. Provided there are a large number of beads, they become evenly distributed.

Viscosity

Water flows more easily than cooking oil, which flows more easily than syrup. The **viscosity** of a fluid is a measure of its resistance to flow. Syrup has a higher viscosity than cooking oil, which has a higher viscosity than water. In general, gases flow much more easily than liquids so the viscosity of a gas is much lower than the viscosity of a liquid.

Viscosity depends on temperature. Oil flows more easily the warmer it is. The lower the temperature of the oil, the less easily it flows. This is part of the reason why starting a vehicle can be difficult on a cold morning. The oil in the engine is cold and therefore less effective as a lubricant. Some liquids become more or less viscous when stirred. For example, paint in a tin becomes 'thinner' if it is stirred enough, which means that it flows more easily and is easier to apply. The opposite effect happens to wallpaper paste when it is stirred. Gradually it becomes 'thicker' and harder to stir until it reaches a consistency allowing it to be spread in a sticky paste onto the wallpaper.

Viscosity is due to the movement of molecules in a fluid between layers moving at different speeds. Molecules that transfer from slow-moving layers to faster-moving layers 'drag' on the faster-moving layers. Thus viscosity is a form of friction between layers in a fluid moving at different speeds. Fast-moving layers in a moving fluid are dragged by slower moving layers, which are dragged by layers moving even slower and so on. For example, when a fluid flows through a pipe, the surface drags on fluid nearby, which drags on fluid further away, which drags on fluid further away and so on.

Pressure

A fluid will generally flow from high pressure to low pressure, driven by the force due to the pressure difference. For example, in most homes, cold water is supplied to a cold water tank via feed pipes connected to a main water pipe outdoors under the ground. Water is pumped into the main pipe at high pressure at a local pumping station and flows along the pipe into the homes when required.

Pressure is caused when an object exerts a force on a fluid. The fluid in a syringe squirts out of the nozzle when a force is applied to the end of the syringe. The greater the force, the faster the fluid squirts out. The narrower the syringe, the easier it is to use, because less force is needed to create a certain pressure with a narrow syringe, compared with a wide syringe. The pressure in the syringe is the force per unit area acting on the fluid. The greater the force, or the smaller the area of the object, the higher the pressure created in the fluid.

Pressure = force per unit area

The unit of pressure is the **pascal** (Pa), which is equal to the pressure exerted by a force of 1 newton acting on an area of 1 square metre, perpendicular to the area.

Hydrostatic pressure

The force of the Earth's gravity causes pressure in a fluid. The pressure in a liquid in an open container increases with depth below the surface. This is why water runs out of a sink more slowly as the sink empties. The pressure of the water at the plughole depends on the depth of water in the sink. As the depth becomes less, the pressure becomes less and the rate of flow drops.

Gas pressure

The pressure of a gas is due to gas molecules hitting the container surface and rebounding. In a gas, the average separation of the molecules is much larger than the size of a molecule. In comparison, the molecules in a liquid or in a solid are always in contact with each other. Gas molecules move about at high speeds, colliding with each other and the container at random. The motion of gas molecules in a box is like the motion of squash balls flying about in a squash court, hitting the walls and rebounding and occasionally hitting each other. The pressure due to the effect of gravity on the gas is much much smaller than the pressure due to the repeated impacts of the molecules on the container. The force due to the impacts is smoothed out because the number of molecules per second hitting the surface of the container is very large.

The first direct evidence for the existence of molecules was obtained by the eighteenth-century Scottish botanist, Robert Brown. He used a microscope to observe tiny pollen grains floating in air and he saw that the grains quivered as they moved about at random. He reasoned that their quivering haphazard motion was because each grain was bombarded unevenly by fast-moving molecules. The uneven impacts by molecules too small to see were forceful enough to push each tiny grain about at random.

The pressure of a gas in a sealed container increases if the gas is heated. This happens because the molecules of the gas move about faster when the gas is heated. Thus the impacts on the container surface are more forceful and more frequent and so the pressure of the gas is greater. If the gas is cooled, the pressure falls as the molecules move more slowly and their impacts are less forceful and less frequent. The pressure would be zero if the gas was cooled to absolute zero, which is −273°C.

The pressure of a gas at constant temperature increases if the volume of the gas container is reduced. For example, if the air in a cycle pump is trapped by blocking the outlet and then compressed by pushing the handle down the barrel, the pressure of the air in the pump increases. The reason why this happens is that the air molecules hit the container surface more often in a smaller space because they have less distance to travel between successive impacts with the container surface.

7

the structure of matter

The ideas about atoms that were first put forward thousands of years ago by Democritus in Ancient Greece were rediscovered by John Dalton in the early nineteenth century. Dalton's theory that atoms are indivisible and indestructible worked well for almost a century until the discovery that atoms contained much lighter negative particles, which became known as electrons. Experiments by Ernest Rutherford in the first decades of the twentieth century showed that every atom contains a positive nucleus where most of its mass is located. Further experiments led to the conclusion that the nucleus is composed of protons and neutrons, and that the electrons move round the nucleus. In this chapter we will look at the discoveries that led to this picture of the atom, before moving on to look at the astonishing picture revealed by physicists probing matter on incredibly small scales. We know now that protons and neutrons are composed of particles called quarks.

Radioactivity

Radioactivity was discovered accidentally by Henri Becquerel when, in Paris in 1896, he was conducting research into the effects of X-rays on uranium compounds. He discovered that these substances exposed to X-rays glow and continued to glow when the X-ray machine was switched off. The substance was described as 'radioactive' because it did not need to be supplied with energy to make it emit radiation and was therefore emitting radiation actively.

The nature of radioactivity was established by Ernest Rutherford, who showed that the radiation is produced when unstable atoms disintegrate.

Radioactivity is due to the instability of a nucleus that has too many protons or neutrons. Such a nucleus becomes stable, or less unstable, by emitting one of three types of radiation, as below. All of these are capable of creating ions in air by knocking electrons out of air molecules.

1 **Alpha radiation (α)** consists of particles, each composed of two protons and two neutrons. An α particle is emitted by a very large unstable nucleus. Alpha radiation:
   * is easily stopped by cardboard or thin metal
   * has a range in air of no more than a few centimetres
   * ionizes air molecules much more strongly than the other two types of radioactive radiation.

2 **Beta radiation (β)** consists of electrons, each emitted when a nucleus with too many neutrons disintegrates. A neutron in such a nucleus suddenly and unexpectedly changes to a proton; in the process, an electron is created and instantly emitted from the nucleus. Beta radiation:
   * is stopped by 5–10 mm of metal
   * has a range in air of about 1 metre
   * ionizes air molecules less strongly than α radiation.

3 **Gamma radiation (γ)** consists of high-energy photons. A photon is a packet of electromagnetic waves. A gamma photon is emitted from a nucleus with surplus energy

after it has emitted an α or a β particle. Gamma radiation:

* is stopped only by several centimetres of lead
* has an infinite range in air
* ionizes air molecules very weakly.

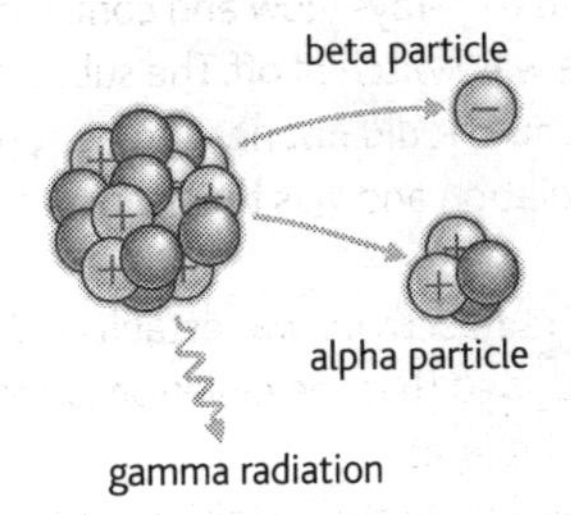

Figure 7.1 *Radioactivity.*

The **Geiger counter** is used to detect radioactive radiation. This device was invented by Hans Geiger, who worked with Rutherford. It consists of a tube connected to an electronic counter. Each time an ionizing particle enters the tube, the counter registers it as a single count and a click is heard. The tube itself is sealed and hollow, with a thin window over one end and a metal rod along its axis. The tube contains gas at very low pressure. With several hundred volts between the rod and the tube, an ionizing particle entering the tube ionizes the gas atoms, which then ionize more atoms, and so on. The gas becomes conducting for a fraction of a second, causing a tiny pulse of electricity to pass through the counter and be registered.

Background radioactivity must be taken into account when measuring the count rate due to a radioactive source. Background radioactivity is caused by radioactive substances present in rocks, by the effect of cosmic radiation on the atmosphere and by radioactive pollutants released into the atmosphere, such as happened at Chernobyl in 1986. To take account of background radioactivity, the count rate is measured with the source present and without the source present. The count rate due to the source is the difference between the two measurements.

Half-life

The **half-life** of a radioactive isotope is the time taken for half the number of atoms of the isotope to disintegrate. Suppose 10 000 atoms of a certain radioactive isotope X are present initially. The number of atoms decreases

* from 10 000 to 5000 after one half-life, then
* from 5000 to 2500 after a further half life, then
* from 2500 to 1250 after a further half life, and so on.

The amount of the radioactive isotope therefore decreases with time as shown in Figure 7.2, which is a **half-life curve**. Half-life values range from a fraction of a second to billions of years. For example, the half-life of uranium 238 ($^{238}_{92}U$) is about 4.5 billion years.

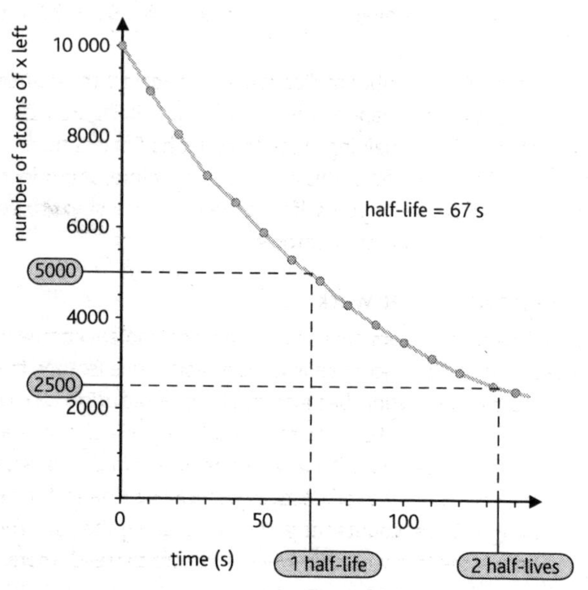

Figure 7.2 *A half-life curve.*

Radioactive disintegration is a random process. For a large number of atoms of a given radioactive isotope, the proportion that disintegrate per second is constant. This follows because of the random nature of radioactive disintegration. To appreciate this, suppose a thousand dice are rolled and all those that show a '1' are then removed. With such a large number to start with, the number of dice removed would be about 167 (= 1000/6) because, on average, one-sixth would show a '1'. If the process was repeated with the 833 remaining dice (= 1000 – 167), the number removed in this second throw would be about 139 (= 833/6). If the process were to be repeated a number of times, each time using the remaining dice, the number remaining would decrease as below.

Number of throws	0	1	2	3	4	5
Number of dice remaining	1000	833	694	578	482	402
Number of dice removed	167	139	116	96	80	67

A graph of the number of dice remaining against the number of throws is the same shape as the half-life curve in Figure 7.2. The number of dice remaining drops to about half in less than four throws. The 'half-life' of this process is, therefore, about four throws. You can prove for yourself that the number would drop to about 250 in about four more throws.

Radioactivity at work

1 **Radioactive tracers** are used in medicine and environmental technology. The tracer needs to be a radioactive isotope that emits β or γ radiation, because this type of radiation can pass through materials. For example, suppose a gas leak occurs in an underground gas pipe. The leak can be pinpointed by injecting a small quantity of a radioactive gas into the pipe and then moving a Geiger counter at ground level along the pipe. The source of the leak would be where the counter reads more than the background count rate.
2 **Radioactive dating** enables the age of ancient objects to be measured. For example, living wood contains a small proportion of a radioactive isotope of carbon, $^{14}_{6}C$. When a tree dies, the proportion of this isotope gradually decreases as these atoms

disintegrate with a half-life of about 5600 years. To measure the age of an ancient wooden object, the count rate due to a sample of the object is measured and compared with the count rate of an equal mass of wood from a living tree. This comparison can then be used to work out how many half-lives have elapsed since the death of the tree used to make the ancient object. If the count rate was reduced to one-quarter, the object would be 2 × 5600 years old (= 2 half-lives).

Radioactive hazards

Radioactivity is dangerous to human health because ionizing radiation kills living cells and causes tumours. For this reason, the use of radioactive sources is subject to strict regulations. For example, radioactive sources must be kept in a lead container to prevent radiation from escaping. Also, radioactive sources should only be moved using a suitable handling device. Radioactive waste from nuclear reactors must be stored in sealed containers at approved sites.

Quarks and leptons

Antimatter was predicted by the British physicist Paul Dirac in 1926. He reckoned that for every type of particle, there is a corresponding antiparticle with identical mass and the opposite type of charge, if the particle is charged. He predicted that a particle and its antiparticle can be created as a pair and they can annihilate each other to produce gamma photons. Within a few years, the American physicist Carl Anderson had detected the positron, the antiparticle of the electron. Anderson first observed the track of a positron in a device known as a cloud chamber. A charged particle passing through a cloud chamber leaves a track because it creates ions along its path. The ions cause tiny droplets to form from the vapour in the chamber, leaving a visible track. Anderson recognized that the track was produced by a particle like the electron, but the magnetic field applied to the chamber made the track curve in the opposite direction to that expected for an electron.

Cosmic radiation consists of particles that enter the Earth's atmosphere from space. These particles are produced by the Sun or other stars. They crash into the nuclei of atoms in the atmosphere, creating cascades of particles at high speed that can reach the Earth's surface. As well as positrons, other particles called pi-mesons (or pions) were discovered. These particles were called mesons because they were found to be 'middle-weight' particles, heavier than electrons but lighter than protons. Further investigations revealed the existence of 'strange' particles that are created in particle pairs, rather than particle–antiparticle pairs, and which decay into pions and protons.

Accelerators have been constructed to create and study these new types of particle and antiparticle. In essence, an accelerator is an evacuated tube containing electrodes, which are used to accelerate charged particles such as electrons or protons to speeds approaching the speed of light. These charged particles are then directed in a narrow beam at a target. Some of the charged particles collide with the target nuclei to create mesons and other short-lived particles and antiparticles from the energy of the charged particles in the beam. Using these accelerators, physicists have discovered a large number of short-lived particles and antiparticles, charged and uncharged, with a range of masses. These newly discovered particles and antiparticles were found to fit patterns that could be explained by assuming that protons and neutrons are composed of three elementary particles, which became known as **quarks**.

The quark model is based on the following assumptions:

1 There are six different types of quark: the up quark, the down quark, the strange quark, the charmed quark, the bottom quark and the top quark (referred to as u, d, s, c, b and t for brevity).

2 For every type of quark, there is a corresponding antiquark.

3 Quarks combine in threes to form particles like the proton and the neutron. Antiquarks also combine in threes to form antiparticles like the antiproton and the antineutron. Such composite particles are collectively referred to as **baryons**.

4 A **meson** consists of a quark and an antiquark.

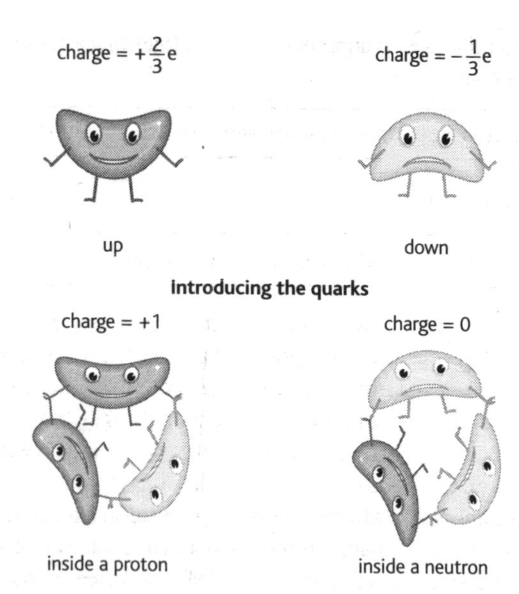

Figure 7.3 *Quarks.*

In terms of the charge of the electron, the u, c and t quarks each carry a charge of $^{+2}/_3\,e$ and the other three quarks carry a charge of $^{-1}/_3\,e$. An antiquark carries an equal and opposite charge to its corresponding quark. The symbol for an antiquark is the same as for a quark but with a bar over the top. For example, $\bar{d}$ (pronounced 'dee bar') represents the symbol for a down antiquark. So:

* A proton is composed of two up quarks and a down quark.
* A neutron consists of an up quark and two down quarks.
* A pion consists of an up or down quark and an up or down antiquark.
* Strange particles contain strange quarks or antiquarks.

The creation of pions and strange particles and antiparticles can be explained using the quark model. For example, if a proton at high speed collides with another proton, the following interaction could take place:

proton + proton → positive pion + proton + neutron

In quark terms:

$uud + uud \rightarrow u\bar{d} + uud + udd$

In the collision, a down quark and a down antiquark are created from the kinetic energy of the high-speed proton. The quarks and the antiquark regroup to form a positive pion, a proton and a neutron.

The quark model was confirmed by physicists using the Stanford Linear Accelerator to accelerate electrons to speeds within a tiny fraction of the speed of light and use them to bombard a target. These electrons were scattered by the target nuclei in directions corresponding to three hard centres in every neutron and proton.

Where do electrons fit into this model? The answer is that they do not. Electrons, positrons and certain other particles and antiparticles are thought to be elementary in the sense that they are not composed of smaller particles. These particles and antiparticles are collectively referred to as **leptons**. All particles not in the quark family belong to the lepton family. Quarks and antiquarks might themselves be composed of even smaller particles. The difference of charge between the up quark and the down quark is equal to the charge of the electron. This perhaps suggests that there is a deeper link between quarks and leptons. The Large Hadron Collider at CERN, the European Centre for Nuclear Research, might provide some answers in due course. This accelerator is designed to collide protons at energies more than 20 times greater than the biggest accelerators in current use. The mechanism linking mass and energy might at last be uncovered, a century after Einstein discovered that mass m and energy E are interchangeable in accordance with the equation $E = mc^2$ where c is the speed of light in free space.

energy from the nucleus

No other branch of science has so far had such an impact on the human race as nuclear physics. After all, as nuclear weapons have the capacity to destroy the human race. Yet before the discovery of nuclear fission, Lord Rutherford reckoned that obtaining energy from the nucleus on a large scale was unrealistic 'moonshine'. Nuclear fission, the splitting of a large nucleus, was discovered by Otto Hahn and Fritz Strassmann in Berlin in 1938. Nuclear reactors for electricity generation were subsequently developed and now account for about 8% of all electricity generation worldwide. Concerns about radioactive waste and the Chernobyl disaster caused nuclear power to fall out of favour, but the current generation of nuclear power stations will need to be replaced if fossil fuel usage is cut back to combat global warming, and electricity cuts and fuel rationing are to be avoided. In this chapter we will look at the principles of nuclear fission, how nuclear reactors work and why they are likely to continue to make a significant contribution to our electricity supplies.

Nuclear fission

A radioactive substance emits radiation because the nuclei of its atoms are unstable and disintegrate at random, emitting α radiation, β radiation or γ radiation. Such changes cause energy release on a scale about a million times greater than when atoms react chemically. However, radioactive isotopes with sufficiently long half-lives are not active enough to release energy at a fast enough rate to produce electricity on a large scale. This is why Rutherford reckoned that nuclear power was unrealistic – until nuclear fission was discovered.

Nuclear fission is the splitting of a large nucleus into two approximately equal fragments. Energy is released in this process on a scale even greater than when radioactive disintegration occurs. Hahn and Strassmann proved that this process happened to the uranium isotope $^{235}_{92}U$. Further investigations showed that this isotope could be made to fission by bombarding it with neutrons, and that two or three neutrons are released when such a nucleus splits. These released neutrons are referred to as fission neutrons. Thus a **chain reaction** is possible in which a neutron splits a uranium 235 nucleus and two or three fission neutrons are released, which go on to split other uranium 235 nuclei, leading to more neutrons being released, which then go on to split more uranium 235 nuclei. Each fission event releases energy and so an enormous amount of energy is released if the chain reaction is maintained.

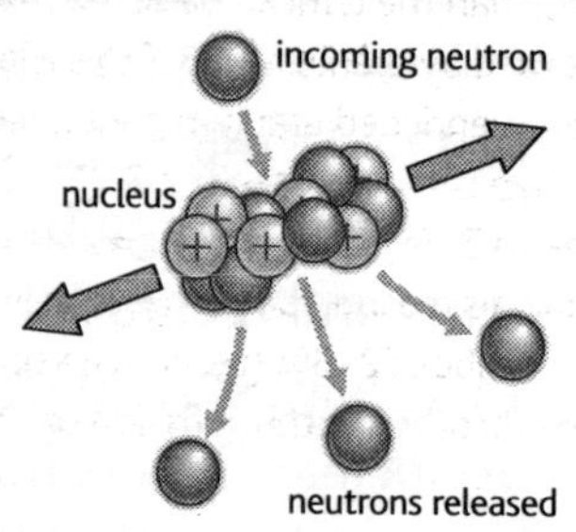

Figure 8.1 *Fission.*

Why should a large unstable nucleus release energy when it fissions or undergoes a radioactive change? Such a nucleus wobbles about and its shape changes as it wobbles. If a neutron from outside hits it when it is distorted into two lobes still joined together, the lobes gain enough energy to separate even more. At sufficiently large separation, the electrostatic repulsion between the two lobes forces them to fly apart. The protons and the neutrons in a nucleus are normally held together by a strong attractive force called the **strong nuclear force** that prevents the protons repelling each other electrostatically due to their 'like' charges. Fission occurs because the strong nuclear force is a short-range force whereas the electrostatic force of repulsion is a long-range force.

Nuclear reactors

A nuclear reactor in normal operation releases energy at a steady rate. This is because the rate of fission events in the reactor is controlled by **control rods** so that each fission event causes exactly one further fission event. The control rods are made of material that absorbs neutrons and they can be moved in or out of the reactor core to keep the number of neutrons in the core constant. Should the fission rate rise, the control rods are pushed into the core a little to mop up surplus neutrons and thus reduce the fission rate.

The mass of uranium 235 must exceed the **critical mass**, which is the least mass for fission to be sustained. If the mass of uranium 235 is less than the critical mass, too many neutrons escape or are absorbed by uranium 238 if this is present. The fuel in a nuclear reactor is enriched uranium contained in sealed fuel rods. Natural uranium is mostly uranium 238 ($^{238}_{92}U$) and it contains less than 1% uranium 235 ($^{235}_{92}U$). Uranium 238 absorbs neutrons without fission. For this reason, the percentage of uranium 235 must be increased to about 2–3% (i.e. enriched); otherwise too many neutrons are absorbed without fission or they escape from the reactor core.

The first atom bomb consisted of two sub-critical hemispheres of pure uranium 235 placed at opposite ends of a hollow cylinder. When the device was detonated, the two hemispheres were forced together into a super-critical mass, which then exploded.

Thermal nuclear reactors

The neutrons released when a uranium 235 nucleus fissions in a nuclear reactor need to be slowed down. This is because, when they are released, they move too fast to cause further fission. The slowing down is achieved by surrounding the fuel rods that contain the uranium 235 with a suitable substance to 'moderate' the speed of the neutrons. This **moderator** needs to be composed of light atoms. The neutrons collide with the moderator atoms and transfer kinetic energy to them in the process. This transfer process is most effective if the moderator atoms are as light as possible. The moderator must also be unreactive. For these reasons, water or graphite (which is carbon) is used as the moderator. After many collisions with moderator atoms, the neutrons move slowly enough to cause further fission. They are then referred to as **thermal neutrons**, because their kinetic energies are of the same order of magnitude as the moderator atoms.

The design of one type of thermal nuclear reactor is shown in Figure 8.2. The fuel rods are placed in channels in the moderator, which is a solid graphite core. The control rods are in channels between the fuel rods. The moderator becomes very hot because its atoms absorb kinetic energy from the neutrons, and so carbon dioxide gas as a coolant fluid is pumped through the moderator channels. The entire core is enclosed in a sealed steel vessel, which is connected via inlet and outlet pipes to a heat exchanger. The coolant is pumped round this sealed circuit and used to raise steam in the heat exchanger to drive electricity generators.

Fuel usage

The complete fission of a kilogram of uranium 235 releases about 8×10^{13} J of energy (= 80 million million joules). A 1000-MW power station operating at an efficiency of 25% needs 4000 million joules of energy from its fuel every second. Each day, the fuel would therefore need to release about 350 million million joules. About 4 kilograms of uranium 235 would therefore be needed to keep a 1000-MW nuclear reactor operating for a day. If the fuel

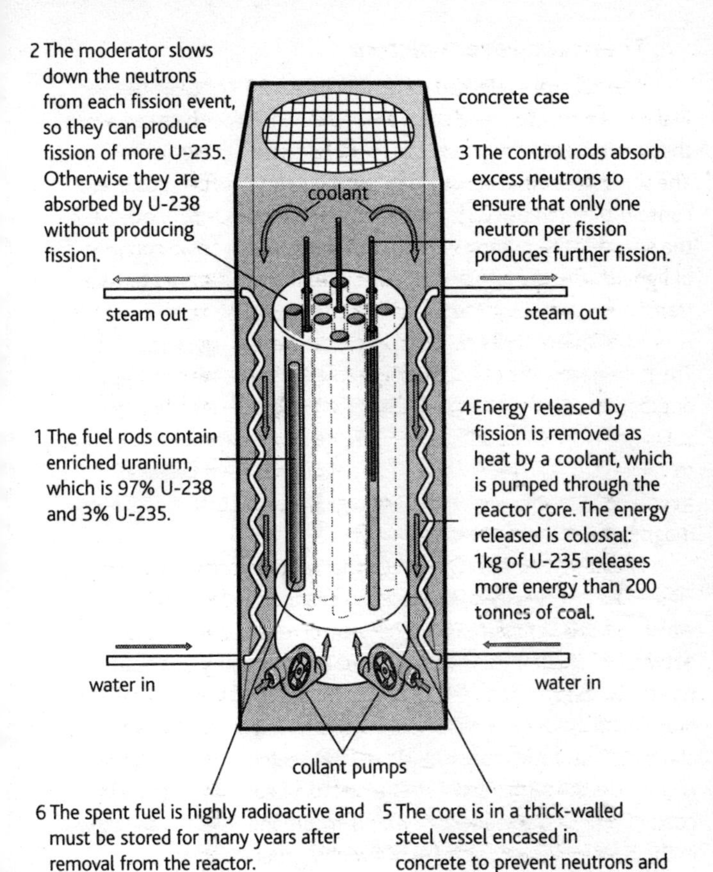

Figure 8.2 *A thermal nuclear reactor.*

for the reactor contained 2% uranium 235, then the mass of fuel needed per day would be about 200 kg (= 0.2 tonnes). In contrast, a 1000-MW oil-fired power station uses about 40 000 tonnes of oil per day.

Notes